A Bond Scheme for Common Agricultural Policy Reform

A Bond Scheme for Common Agricultural Policy Reform

Edited by

Alan Swinbank and Richard Tranter

Centre for Agricultural Strategy, The University of Reading, UK

CABI Publishing

CABI Publishing is a division of CAB International

CABI Publishing
CAB International
Wallingford
Oxfordshire OX10 8DE
UK

Tel: +44 (0)1491 832111
Fax: +44 (0)1491 833508
E-mail: cabi@cabi.org
Website: www.cabi-publishing.org

CABI Publishing
875 Massachusetts Avenue
7th Floor
Cambridge, MA 02139
USA

Tel: +1 617 395 4056
Fax: +1 617 354 6875
E-mail: cabi-nao@cabi.org

A catalogue record for this book is available from the British Library, London, UK.

Library of Congress Cataloging-in-Publication Data
Swinbank, A. (Alan)
A bond scheme for common agricultural policy reform / Alan Swinbank and Richard Tranter.
p. cm.
Includes bibliographical references and index.
ISBN 0-85199-744-9 (alk. paper)
1. Agriculture and state--European Union countries. 2. Governement securities--European Union countries. I. Tranter, R. B. II. Title.

HD1917.75.S956 2004
338.1'3--dc22

2004007863

ISBN 0 85199 744 9

Printed and bound in the UK by Cromwell Press, Trowbridge, from copy supplied by the author.

Contents

Contributors

Leonardo Costa: Faculty of Economics and Management, The Portuguese Catholic University, Centre Regional do Porto, Rua Diogo Botelho 1327, 4169-005 Porto, Portugal

Arlindo Cunha: Praceta das Caleiras 47, 4470-155 Maia, Portugal

Carsten Daugbjerg: Department of Political Science, University of Aarhus, Universitetsparken, DK-8000 Aarhus C, Denmark

Thomas Knapp: Institut für Agrarökonomie, Georg-August Universität Göttingen, Platz der Göttinger Sieben 5, 37073 Göttingen, Germany

Jonathan Little: Centre for Agricultural Strategy, School of Agriculture, Policy and Development, The University of Reading, PO Box 237, Earley Gate, Reading, Berkshire, RG6 6AR, UK

Américo Mendes: Faculty of Economics and Management, The Portuguese Catholic University, Centre Regional do Porto, Rua Diogo Botelho 1327, 4169-005 Porto, Portugal

Miguel Sottomayor: Faculty of Economics and Management, The Portuguese Catholic University, Centre Regional do Porto, Rua Diogo Botelho 1327, 4169-005 Porto, Portugal

Alan Swinbank: Centre for Agricultural Strategy, School of Agriculture, Policy and Development, The University of Reading, PO Box 237, Earley Gate, Reading, Berkshire, RG6 6AR, UK

Stefan Tangermann: Directorate for Food, Agriculture and Fisheries, OECD, 2 rue André Pascal, F-75775 Paris Cedex 16, France.

Richard Tranter: Centre for Agricultural Strategy, School of Agriculture, Policy and Development, The University of Reading, PO Box 237, Earley Gate, Reading, Berkshire, RG6 6AR, UK

Preface

A succession of agricultural economists, way back to the 1960s, has advocated the use of decoupled compensation payments to facilitate reform of farm policies. In Chapter 3, we outline some of the issues and concerns revealed by that literature. In particular, in 1991, a report commissioned by the Land Use and Food Policy Inter-Group (LUFPIG) of the European Parliament proposed a significant development of the concept of decoupled payments: a bond scheme in which the rights to future compensation payments would be vested in a paper asset - the bond - that could then be retained by the original recipient or sold on the stock exchange (see, in particular, Tangermann, 1991). Although the European Commission did propose a transferable bond as part of its reform proposals for milk in 1991, and the Danish Government pressed for a bond scheme rather than arable area payments in the debate over the 1992 CAP reforms, the bond scheme was not adopted at that time. Chapter 6 assesses why it was that the Danish Government's proposals were ill-fated.

In the event, the 1992 reforms introduced arable area payments into the CAP, and modified (and for beef increased) the headage payment schemes for beef cattle and sheep. Although the new arable area payments were introduced to *compensate* farmers for the revenue loss, a decade later the European Commission admitted that:

> 'they have lost part of their compensatory character after ten years of implementation and have instead become simple direct income payments. Therefore, the term 'direct aid' has replaced 'compensation payment'' (Commission, 2002a).

The 1992 (MacSharry) reforms did, however, facilitate the conclusion of the Uruguay Round of GATT negotiations, which made special provision for 'direct payments under production limiting programmes', so-called blue box payments, into which category the EU's area and headage payments fell. The CAP reforms of Agenda 2000 further embedded the concept of direct payments into the CAP - introducing them for milk producers from 2005 for example - although it was now expected that the Member States would introduce an element of 'cross compliance' into their payment regimes, under which farmers would be obliged to respect specified environmental standards if they were to receive payment in full. However, with the prospect of EU enlargement to embrace up to ten candidate states from Central and Eastern Europe (the CEECs), and three Mediterranean countries, and a new round of WTO trade negotiations about to begin (since launched as the Doha Development Agenda), doubts were raised about the future of direct payments. These developments are outlined, and concerns explored, in Chapter 2.

It was against this background that a research project under the EU's Fifth Framework Programme, to re-examine the feasibility and practicality of introducing a bond scheme as an element of CAP reform, was proposed to DG Research and accepted (QLRT-1999-01510). The project involved research teams from The University of Reading (UK), the Georg-August Universität, Göttingen (Germany), and The Portuguese Catholic University (Portugal). We are grateful to the European Commission for this financial support, and to our desk officer Dr. Hans-Joerg Lutzeyer for his support.

The Research Team was:
The University of Reading
Alan Swinbank, Jonathan Little and Richard Tranter
Georg-August Universität Göttingen
Stefan Tangermann, until 31 January 2002, and Thomas Knapp
The Portuguese Catholic University
Américo Carvalho Mendes, Leonardo Costa, Miguel Sottomayor
with, as associates:
Arlindo Cunha, MEP, until October 2003
Carsten Daugbjerg, University of Aarhus

In revisiting the bond scheme proposal, an initial task was to re-articulate its component parts and clarify its scope. As Chapter 6 makes clear, in the Council discussions of late 1991 and early 1992, there was some confusion over the scope of the proposed scheme. Thus, an important part of this project was to clearly articulate the scope and intent of the proposal. This was done in Swinbank and Tangermann (2000 and 2001), as set out in Chapter 4. It is important to note that what was proposed was a transformation of existing direct payments (i.e. the area and headage payments of the MacSharry reforms) into the bond scheme: thus all other aspects of the CAP - the residual elements of price support for cereals and beef, the full support arrangements for other commodities, and the second pillar - would remain in place. By contrast, Beard and Swinbank (2001) had advocated a bond scheme to displace *all* price and income support elements of the CAP.

Swinbank and Tangermann (2001) also noted that, although the purpose of their paper was to argue the merits of a fully decoupled system of compensation (or adjustment) payments, as exemplified by the bond scheme, they saw this 'embedded in a comprehensive reform of the CAP to better reflect the public's wider concerns about the environment and rural development'. This had already led to considerable debate within the research team, with some members arguing that it would be difficult to assess the bond scheme in isolation: that it could only be judged in the context of that wider reform; and that it was beholden on the team to flesh out more details. The consultation phase, on the whole, endorsed these concerns, although the vast majority of the 5,950 British, German and Portuguese farmers who responded to our survey seemed to have no difficulty in grappling with the concept. Nonetheless, in Chapter 5, we explain in more detail how the bond scheme would facilitate transformation of the present CAP into something akin to the Common Agricultural and Rural Policy for Europe (CARPE) advocated by Buckwell *et al.* (1997).

The consultation phase involved discussions with, or written communications from, stakeholder groups in Germany, Portugal and the UK, including environmental, farm, landowner, food industry and consumer groups, as well as government officials and fellow academics. A workshop in Porto, under Chatham House rules, further advanced our thoughts. We are extremely grateful to all our interlocutors, whose insights and reflections permeate this book. Chapter 7 in particular addresses implementation issues and outlines stakeholder concerns.

In July 2002 the European Commission tabled its proposals for a mid-term review of Agenda 2000 (Commission, 2002b): a debate concluded by the Council decisions of June 2003. This is discussed in Chapters 2 and 9. The *Fischler Reforms* of 2003 have further decoupled support from agricultural production, although the new Single Payment Scheme does not embrace all the features of a bond scheme. The survey of farmers' intentions undertaken in Germany, Portugal and the UK, reported in Chapter 8, gives useful insights into how farmers might respond not only to the bond scheme, but also to the new Single Payment Scheme.

Interestingly, we found that, whichever country the responding farmers were from, their responses were very similar. Some 30% of the respondents in the UK, Germany or Portugal said they would alter their mix of farm activities, or enterprises, when support payments are decoupled from current land use or production. However, in Portugal, compared to the UK and Germany, proportionately more of their farmers would take up an entirely new enterprise, with forestry and woodland being the most common choice.

Two noticeable differences between the three study countries were found though. First, in relation to land abandonment, once decoupling had happened and the bond was available, whilst 50% in each country said they would not idle any land, and many would idle less than half their land, in Portugal around 20% would idle all or more than half their land, a much larger proportion than in Germany or the UK. Second, when considering whether farmers would cash in their bonds on the financial market, whilst relatively few in each country said they would do nothing, German farmers were especially against this in comparison with British and Portuguese farmers. Furthermore, when receiving the bond, German farmers were much more unlikely to diversify their farm business than their counterparts in Portugal and the UK.

At the practical farm level, our survey responses and those from the consultation phase, have indicated that there would be relatively few problems if the bond scheme were implemented.

Chapters 9 and 10 review the pressures that continue to bear upon the CAP. Notwithstanding the considerable political opposition to a full decoupling of direct payments, we conclude that a bond scheme could have a useful, and practical, role to play in future CAP reform.

We are extremely grateful to all our colleagues for bearing patiently our editorial whims, and to Teresa Hicks for her skilful preparation of camera-ready copy. We are also grateful to Freda Miller for compiling the index and to John Davis, editor of *EuroChoices*, for his permission to reproduce in Chapter 4 material that first appeared in that publication. Neither the European Commission, nor any of the institutions for which the authors work, have any responsibility for the accuracy of information contained, or views expressed, in this text.

Alan Swinbank and Richard Tranter
Centre for Agricultural Strategy, The University of Reading, UK
June 2004

References

Beard, N. and Swinbank, A. (2001) Decoupled payments to facilitate CAP reform. *Food Policy* 26, 121-145.

Buckwell, A., Blom, J., Commins, P., Hervieu, B., Hofreither, M., von Meyer, H., Rabinowicz, E., Sotte, F. and Sumpsi Vîna, J.M. (1997) Towards a common agricultural and rural policy for Europe. *European Economy Reports and Studies No 5.* CEC, Luxembourg.

Commission of the European Communities (2002a) *Enlargement and Agriculture: Successfully Integrating the New Member States into the CAP. Issues Paper.* SEC(2002)95. CEC, Brussels.

Commission of the European Communities (2002b) *Communication from the Commission to the Council and the European Parliament. Mid-term Review of the Common Agricultural Policy.* COM(2002)394. CEC, Brussels.

Swinbank, A. and Tangermann, S. (2000) *A Proposal for the Future of Direct Payments Under the CAP.* Working Paper No 1 of the Bond Scheme Project. Centre for Agricultural Strategy, The University of Reading, Reading, UK.

Swinbank, A. and Tangermann, S. (2001) The future of direct payments under the CAP: a proposal. *EuroChoices* Premier Issue, Spring, 28-29, 32-34.

Tangermann, S. (1991) A bond scheme for supporting farm incomes. In: Marsh, J.S., Green, B., Kearney, B., Mahé, L., Tangermann, S. and Tarditi, S. (eds) *The Changing Role of the Common Agricultural Policy: the Future of Farming in Europe.* Belhaven, London.

List of Abbreviations

AAPS	Arable area payment scheme: i.e. area payments
ACP	African, Caribbean and Pacific
AMS	Aggregate Measurement of Support
AMTA	Agricultural Market Transition Assistance
AWU	Agricultural Work Unit
BMPs	Best Management Practices
CAP	Common Agricultural Policy
CARPE	Common Agricultural and Rural Policy for Europe; as advocated by Buckwell *et al.* (1997) - see Chapter 4
CEC	Commission of the European Communities
CEECs	Central and Eastern European Countries: ten applicants for EU membership (Bulgaria, the Czech Republic, Estonia, Hungary, Latvia, Lithuania, Poland, Romania, Slovakia, Slovenia)
COPA	Comité des Organisations Professionnelles Agricoles
DG VI	Directorate General for Agriculture within the European Commission
EAGGF	European Agricultural Guidance and Guarantee Fund
EARNs	Euro Area Reference Note, issued by the EIB
ECLP	Environmental and cultural landscape payments
EIB	European Investment Bank
EMU	Economic and monetary union
EU	European Union: the Maastricht Treaty on European Union (which entered into force on 1 November 1993) created an EU consisting of three pillars: the European Communities (including the old European Economic Community - EEC - dating back until 1958, and providing the legal footing for the CAP), a common foreign and security policy and co-operation in the field of justice and home affairs
EU-15	The EU from 1995, comprising 15 Member States
EWG	Environmental Working Group
FAIR	Federal Agricultural Improvement and Reform Act
FEOGA	Funds Européen d'Orientation et de Garantie Agricole: see EAGGF
FSRIA	Farm Security and Rural Investment Act
GATT	General Agreement on Tariffs and Trade
IACS	Integrated Administration and Control System

LFA	Less Favoured Area
LUFPIG	Land Use and Food Policy Inter-Group of the European Parliament
MS	Market stabilisation
MTR	Mid-term review of Agenda 2000, launched by the Commission in July 2002
NAFTA	North American Free Trade Area
NVApp	Net value added at parity
OECD	Organisation for Economic Co-operation and Development
PEG	Production entitlement guarantee; as advocated in Blandford, D. *et al.* (1989) - see, for example, Chapter 4
PFC	Production flexibility contract (also called Agricultural Market Transition Assistance (AMTA)) payments in the US under the 1996 *Federal Agricultural Improvement and Reform* (FAIR) *Act*
PPP	Polluter Pays Principle
PROCAMPO	Programa de Apoyos Directos al Campo, in Mexico
PSE	Producer Subsidy Equivalent until 1998 when the OECD changed it to Producer Support Estimate
RDI	Rural Development Incentive
SCA	Special Committee on Agriculture, which serves the Council of the European Union
SPS	Single Payment Scheme
TAA	Transitional Adjustment Assistance
UAA	Utilised agricultural area
URAA	Uruguay Round *Agreement on Agriculture*
WTO	World Trade Organization

Chapter One:

Decoupled Payments and a Triad of Policy Objectives: Compensation, Farm Income Support and Multifunctionality

Alan Swinbank
Centre for Agricultural Strategy, The University of Reading, UK

The purpose of the study reported here is to explore the feasibility and practicality of deploying a particular form of decoupled payment: the bond scheme. Under the proposal, the CAP's existing area and headage payments would be fully decoupled from production. This means that farmers would no longer have to sow crops or keep livestock to qualify for the annual payments, or even farm their land. Each recipient would then have a guaranteed future stream of payments and, if the right to receive payments - guaranteed by the EU or the Member States - could be transferred by gift, inheritance or sale, the recipient would, in effect, have acquired a paper asset (akin to a government bond) that could at any time be sold on the stock exchange to raise a capital sum reflecting the present day valuation of that future income stream. This, it is argued, could facilitate CAP reform.

This is not a new idea, and indeed the suggestion that decoupled compensation payments have a role to play in farm policy reform dates back to at least the 1960s, as documented in Chapter 3. However, an examination of the literature and the past policy debate suggests that a number of different themes, and conflicting policy objectives, can be confused; hence there is a danger that the purpose of a bond scheme might be misunderstood.

It is perhaps helpful to suggest that there are three themes that can be found in the literature (Fig. 1.1). Sometimes decoupled payments have been advocated as compensation for the removal of price or income support policies, facilitating adjustment by the farm sector to a new policy environment. Almost by definition, such schemes are temporary in nature. This is illustrated by the point at the apex of the triangle in Fig. 1.1. The bond scheme explored in this book occupies this space in the triangle.

Other authors, whilst advocating reform of existing price and income support policies, have suggested that these policies need to be replaced by decoupled income support payments that would be particularly beneficial to low income farmers: a sort of social security scheme for the farm sector. Indeed, Fennell

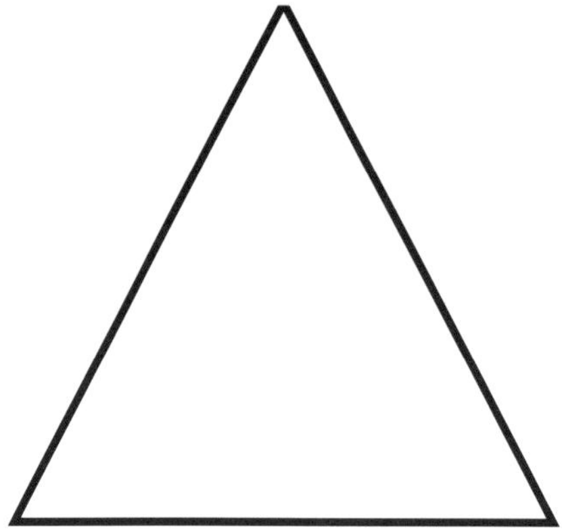

Fig. 1.1. Decoupled payments and a triad of policy objectives

(1997) suggests that this was one of the original aims of the CAP that has long been forgotten in the formulation of policy. Almost by definition such schemes would be a permanent feature of the CAP. This position is represented by the right-hand corner at the base of the triangle in Fig. 1.1. As recently as 1987 the Commission proposed decoupled income support along these lines. In the context of a 'rigorous policy as regards pricing', the Commission concluded that there needed to be 'an additional instrument for income support' that would 'shield certain types of farmers against the impoverishment and hardship which certain adjustments could well force on them'. But at the same time, 'there must be no doubt that the aid is related to the farmer's income and not the activity of his enterprise: thus it would be quite wrong to link the amount paid either directly or indirectly to production levels, prices or inputs and facilities used' (Commission, 1988). A limited scheme was eventually agreed in 1989 (Fennell, 1997).

The left-hand base corner of Fig. 1.1 represents a more recent idea: that farmers, as the custodians of the countryside, should receive payments that are decoupled from the production of crops and livestock, but paid to reward them for the provision of non-marketable benefits that the rest of society values (such as landscape, diversity of flora and fauna, cultural heritage, food security, animal welfare, etc.). This strand of thought was given added policy significance by the conclusion of the Uruguay Round of GATT negotiations. First, the determination of a list of decoupled policies (listed in Annex 2 to the *Agreement on Agriculture*, which is known as the 'green box') and, second, the agreement that 'non-trade concerns' would be taken into account in the further round of multilateral trade negotiations foreshadowed in Article 20, addressing 'the long-term objective of substantial progressive reductions in support and protection'. (These issues will be outlined in detail in Chapter 2.) In the EU these thoughts are now encapsulated in the term 'multifunctionality', and there is a growing literature (and policy debate) over the extent to which the multifunctional facets of farming can, or cannot, be unbundled from farm output. We return to this discussion in Chapter 5.

In practice, much of the literature, and a number of policy proposals, occupy space *within* the triangle depicted in Fig. 1.1, and clarification is frequently required with regard to the authors' or proponents' use of the term 'decoupled' and of the policy objectives addressed. As we shall outline in subsequent chapters, the CAP's area and headage payments, introduced in 1992 to *compensate* farmers for policy reform, whilst not fully decoupled from production, have over time come to be seen as offering income support to farmers and they are sometimes justified as a mechanism for rewarding farmers for their supply of the multifunctional characteristics of agriculture. In the next chapter we introduce the 1992 reforms and direct payments.

References

Commission of the European Communities (1988) *The Agricultural Situation in the Community. 1987 Report*. Office for Official Publications of the European Communities, Luxembourg.

Fennell, R. (1997) *The Common Agricultural Policy. Continuity and Change*. Clarendon Press, Oxford, UK.

Chapter Two:

Direct Payments in the EU and their Treatment in the WTO

Alan Swinbank
Centre for Agricultural Strategy, The University of Reading, UK

The bond scheme proposed in Chapter 4 would apply to existing direct payments under the CAP, and in particular to the arable area payments introduced by the so-called MacSharry reforms of 1992. The Fischler reforms of 2003 have begun, but not completed, the transition to a bond scheme. The purpose of this chapter is to document the importance of direct payments in the EU, and to explain their treatment in the WTO.

The origins of the 1992 reforms

Kay (1998) suggests that the *genesis* of the MacSharry reforms of May 1992 lay in the growing realisation on the part of Commission President Jacques Delors that the 'stabiliser' package of February 1988 would not suffice to bring CAP spending under control, and consequently that Delors' ambitious plans for the deepening of European integration could be jeopardised. Thus, under the second Delors Commission (1989-1992), the incoming Irish Commissioner, Ray MacSharry, was specifically chosen by Delors to take on responsibility for agriculture; and from January 1989 a small team headed by MacSharry began work on the next CAP reform. The team worked in secret, and 'It is clear that few of the commodity division heads of DG VI (now known as DG Agriculture) knew that reform proposals were being prepared in 1989/90' (Kay, 1998). Ross (1995) too, reports on the existence of a 'reform' team, but with a somewhat different emphasis to that of Kay: work had begun in 'early 1990' after Delors had set out 'initial lines of thought', with a brainstorming in June, and the 'conceptualisation' of ideas in September 1990. Kay's account emphasises MacSharry's role and internal pressures for reform; Ross that of Delors; but both agree that Delors and MacSharry had formulated their ideas for CAP reform by the time the Uruguay Round discussions collapsed in December 1990 (Ross, 1995).

Throughout the 1980s there had been various attempts to reform the CAP, and control budget expenditure. However, despite the introduction of 'guarantee thresholds' in 1982 - which were supposed to trigger reductions in price guarantees if production exceeded predetermined thresholds - and milk quotas in 1984, budget expenditure continued to soar (Swinbank and Tanner, 1996). The British Prime Minister, Margaret Thatcher, had been resisting an agreement to increase the EU's budgetary resources to fund the widening range of EU policies until the CAP was reformed. The package deal that emerged, at a meeting of the European Council in February 1988, both expanded the EU's budgetary resources and suggested that a fundamental reform of the CAP had been agreed. The arrangements were complex: automatic 'agricultural stabilisers' would come into play if the annual growth in budget expenditure on CAP price support exceeded 74% of the increase in the EU's gross domestic product (GDP), with the Council ceding responsibility to the Commission to reduce support prices if commodity specific thresholds were surpassed (Swinbank and Tanner, 1996).

The Uruguay Round of multilateral trade negotiations under the auspices of the General Agreement on Tariffs and Trade (GATT) was also under way. Launched in September 1986, the Punta del Este Declaration - that had only been accepted by the EU after considerable pain and deliberation - talked of achieving 'greater liberalisation of trade in agriculture' and of bringing 'all measures affecting import access and export competition under strengthened and more operationally effective GATT rules and disciplines' (Swinbank and Tanner, 1996). How the EU was to live up to that promise was unclear, and remained so when MacSharry took over as Commissioner for Agriculture in January 1989. Indeed, a 'mid-term' review of progress in the Uruguay Round, held in Montreal in December 1988, in revealing *inter alia* the enormous gulf between participants on the issue of liberalisation of trade in agricultural products, had just about brought the whole process to a halt (Josling *et al.*, 1996).

Various authors have outlined the difficulties the EU encountered in the closing months of 1990 as it strove to agree a negotiating mandate on agriculture in, first, the Commission, and then the Council, in preparation for what was scheduled to be the final Ministerial Meeting of the Uruguay Round (see, for example, Swinbank and Tanner, 1996; Kay, 1998). The failure of that meeting at the Heysel conference facility in Brussels in December 1990 is seen by Ross (1995) to be 'a disguised blessing': the inability of GATT Ministers to agree on agriculture was a 'convenient shield' that hid the fact that negotiations on other issues were still in doubt, and 'For Delors and his team, the collapse granted precious time to get CAP reform through the Commission and then win Council approval'. Similarly, Kay (1998) suggests that MacSharry and his team simply had not had enough time since January 1989 to bring forward plans for CAP reform in advance of the EU's determination of its negotiating position in the GATT round. Tangermann (1998) suggests that MacSharry engineered the collapse of the agriculture negotiations at Heysel, because of MacSharry's belief that, to be politically acceptable, CAP reform had to come before the GATT agreement, thus *facilitating* that agreement, rather than after the agreement, *forced* on the EU by the international community.

Whatever the motives, roles and strategies of the participants, it would appear that the key elements of the MacSharry reform had been thought out by December 1990. In October 1990, in trying to secure agreement on an EU 'offer' in the Uruguay Round negotiations, the Commission revealed the outline of the proposed MacSharry reform (Agra Europe, 1990; Swinbank and Tanner, 1996). Kay (1998) reports meetings of key players from DG VI to advance the plans in the week following the Heysel impasse, and in January 1991, the College of Commissioners was presented with the plan at an informal Sunday seminar (Ross, 1995).

The proposals had been formulated over a long period, but without the leverage of the GATT negotiations in which the EU's wider economic interests forced change on the agricultural sector, it seems unlikely that the 1992 reforms would have been approved by Farm Ministers.

The 1992 (MacSharry) and 1999 (Agenda 2000) reforms

The 1992 CAP reforms formed a complex package, covering not only the support mechanisms for individual commodities, but also various 'accompanying measures' (Commission, 1993; Swinbank, 1993, 1997). For the purpose of this study, it is important to note that what emerged was 'The introduction of a system of permanent compensatory aid to neutralize the negative effect on incomes caused by the decision to lower prices in the cereals, oilseeds and beef and veal sectors' (Commission, 1993). It should be noted that the Commission referred to *compensatory* aid, but that this aid was said to be *permanent*.

For cereals there was a significant cut in the intervention price, in three annual steps, from an average buying-in price of 177.49 ecu per tonne in the period immediately preceding the MacSharry reforms, to 119.19 ecu per tonne from 1995 on. Table 2.1 details the evolution of prices. To compensate for this fall in support prices, a direct payment was progressively introduced, reaching a rate of 54.34 ecu per tonne from 1995. The Council Regulation of July 1992, establishing this new system of support, referred to the need to 'compensate for the loss of income caused by the reduction of the institutional prices by a compensatory payment to those who sow such products' (Council, 1992). The Agenda 2000 reforms, of March 1999, saw a further 15% cut in the intervention price, and an increase in the direct payment to compensate for half this price cut (see Ackrill, 2000 or Swinbank, 1999 on the Agenda 2000 reforms).

The compensatory aid is paid as an area payment. Member States had to specify one or more regions, and then determine the average yield of cereal crops in each region for the period 1986/87 to 1990/91, excluding the years with the highest and lowest yields. The UK, for example, declared five regions, with an average yield in England of 5.93 tonnes per hectare. The area payment in any particular region is the basic amount of aid (from Table 2.1) multiplied by the 'regional reference yield'. As a consequence, area payments differ markedly from one region of the EU to another, reflecting climatic and topographical differences, and the level of technology applied in cereal production at the beginning of the 1990s. As part of the 1999 reforms, both Italy and Spain secured an upward revision in their reference yields (Council, 1999a).

Table 2.1. Support prices for cereals following the 1992 and 1999 reforms

	ecu or €/tonne	
	Intervention Price (July)	Aid (Basic Amount)
Pre-1992	177.49 *	—
1995/96 to 1999/2000	119.19	54.34
2000/01	110.25	58.67†
2001/02 on	101.31	63.00†

* an average buying-in price of 155 ecu per tonne, multiplied by the correcting factor of 1.145109 valid at the time of the 1992 reforms. From April 1984 until February 1995, as a consequence of complications under the green money system, a so-called 'switch-over mechanism' applied that meant that a 'correcting factor' had to be applied to intervention and other support prices. The system was abolished on 1 February 1995, and all support prices, expressed in ecu, were automatically increased by the correcting factor then in force (Swinbank and Tanner, 1996).

† plus €19 per tonne in Finland and the arctic zones of Sweden.

An additional area aid is paid to durum wheat producers: at €344.5 per hectare in 'traditional' production areas, and at €38.9 per hectare in 'non-traditional', but nonetheless defined, regions. As a result of the Agenda 2000 reforms, producers in Finland and northern regions of Sweden receive an augmented payment (of €19 per tonne, multiplied by the regional yield), to reflect the additional drying costs they incur (Council, 1999a).

Oilseeds and protein crops

During the Dillon Round of GATT negotiations the EU had bound its import duties on soybeans and other oilseeds at a zero tariff. Subsequently it introduced a scheme under which it paid a subsidy to crushers when they processed oilseeds (mainly rape and sunflower) sourced from EU producers. The US challenged these provisions in the GATT, and a GATT panel agreed that the EU discriminated against imported products. In an attempt to produce a GATT-compatible policy, in 1992 the EU introduced an area payment system; but, following further complaint from the US, the original GATT panel had been reconvened earlier in the year. Although the reconvened panel agreed that the EU system was now compatible with GATT rules, it nonetheless concluded that the policy still had the effect of impairing the tariff concessions granted to the US in the Dillon Round. Thus the EU had the option of making further changes to its oilseeds policy, or seeking a negotiated outcome with the US (Commission, 1993). At Blair House, the US President's official guest house in Washington, DC, in November 1992, a compromise was reached that ultimately became part of the Uruguay Round accords (Swinbank and Tanner, 1996). In the meantime, in the MacSharry package, it had been agreed that the area payment scheme for oilseeds would become part of the area payment scheme for arable crops, albeit at a higher level of payment than that for cereals. As a result of Agenda 2000 these higher area payments on oilseeds have been phased down, and from 2002 the area payment on oilseeds is the same as that for cereals. Linseed, grown either for flax or oil, is now also covered by the scheme.

Similarly, from 1993, protein crops (peas, field beans and sweet lupins) were covered by the area payment scheme, but again with higher payments than those applicable for cereals. And in Member States where maize silage is not a 'traditional crop', Agenda 2000 introduced the possibility of making area payments on grass silage.

Set-aside, IACS, the simplified scheme and base areas

Farmers must fulfil certain conditions before they can claim arable area payments. They must make their claims through the Integrated Administration and Control System (IACS) that also serves as a validating procedure to counter fraudulent applications and monitor the total area claimed (House of Commons, 2001). Claims can only be made on land that was in eligible uses (basically land in an arable rotation, and earlier set-aside programmes) on 31 December 1991, and cannot be claimed on land which is declared as 'forage area' under the beef premia schemes. Consequently the national IACS data bases now have an almost complete record of land use in agriculture.

Claims can only be made on eligible crops, and a normal crop must be sown and maintained until flowering, although claimants are not obliged to harvest their crop. Most claimants are obliged to set-aside a specified percentage of their arable land. This percentage can be varied on an annual basis, but it currently stands at a default rate of 10%. Thus, to claim arable area payments on 90 hectares of crops, a farmer must set-aside (i.e. not grow crops eligible for CAP price support) a further 10 hectares of eligible land. Additional land can be set-aside on a voluntary basis. Complex rules exist to determine what can, or cannot, be done on set-aside land, with key dates determining applicable periods: for example, after 15 July you can prepare and sow crops on set-aside land for harvest the following year (MAFF, 1999). Under certain restrictive circumstances, so-called 'non-food' crops can be grown on set-aside land; what this means is that CAP price support does not apply, and the crops are used for industrial processing.

Arable area payments are also paid on set-aside land. Initially payment was at the same level as that for cereals, but payments were then increased by the Council as it attempted to assuage French concerns about the negative impact of the Blair House Accord on EU agriculture (Swinbank and Tanner, 1996). Following the Agenda 2000 reforms, arable area payments on set-aside land are now re-harmonised on the cereals rate.

However, if farmers are only claiming for a relatively small area of arable area payments, they are excused the set-aside requirement. The area involved is determined on a regional basis, using the 'regional reference yield', and equates to the production of 92 tonnes of cereals. This averages out at about 20 hectares across the EU. In 1998, 67% of the EU-15 applicants for arable area aid took advantage of this, accounting for 20% of the total area claimed: down markedly from the figures in the first year of the scheme (see Table 2.2).

Table 2.2. Applications for arable area aid under the simplified scheme

Simplified scheme applications as a percentage of all applications:	% by area	% by number of applications
1993/94	31.7	77.9
1994/95	28.6	76.2
1995/96	27.1	73.1
1996/97	25.7	72.4
1997/98	23.9	67.9
1998/99	20.0	66.6

Source: European Commission (2001a)

In addition to the requirement that arable area payments (on both cropped and set-aside land) can only be claimed on land in an arable rotation in December 1991, Member States have had to declare a *base area* for each region, and - within that - a *maximum guaranteed area* for oilseeds. If the total claims in any region exceed the base area, then all claims are scaled back proportionally. This has proved problematic for the Eastern Länder of Germany, but one outcome of the Agenda 2000 reforms was an increase in their base area by 150,000 hectares (Council, 1999a). The separate, and additional, limitation on oilseeds stems from the Blair House Accord which set limits on the area of oilseeds that could benefit from area payments (Swinbank and Tanner, 1996). Unlike the base area limitations, the oilseed limits have had a real impact on support. Thus in 1998, the maximum guaranteed area was overshot by 8.4%, triggering country-specific reductions in support payments ranging as high as 21.3% in Ireland (European Commission, 2001a). The EU is of the view that, with the harmonisation of area payments on oilseeds to the rate received by cereal producers as a result of the Agenda 2000 reforms, oilseed producers will no longer receive *crop specific* payments, and the Blair House limitations will lapse. This view is contested by the US.

Soaring budget costs

In the early 1990s, prior to the policy changes brought about by the MacSharry reforms, EU budget expenditure on support of cereals, oilseeds and protein crops taken together amounted to about 9.7 billion ecu per year, some 31.5% of European Agricultural Guidance and Guarantee Fund (EAGGF) guarantee expenditure (three year average, 1990-1992; Commission, 1993). (Although a separate fund no longer exists, DG Agriculture still refers to the EAGGF in allocating budget expenditure. Over the decade the EU expanded from 12 to 15 Member States, and there has been some change in the allocation of expenditure between the Guidance and Guarantee Funds of EAGGF, but these details do not change the overall picture reported in this paragraph.) A decade later, in 2001, expenditure had nearly doubled, to €17.5 billion, accounting for over 40% of EAGGF guarantee expenditure (Table 3.4.3.1, DG Agriculture web site). The change is shown in Fig. 2.1, and clearly indicates that the introduction of arable area aid both increased, and changed the structure of, expenditure. Although

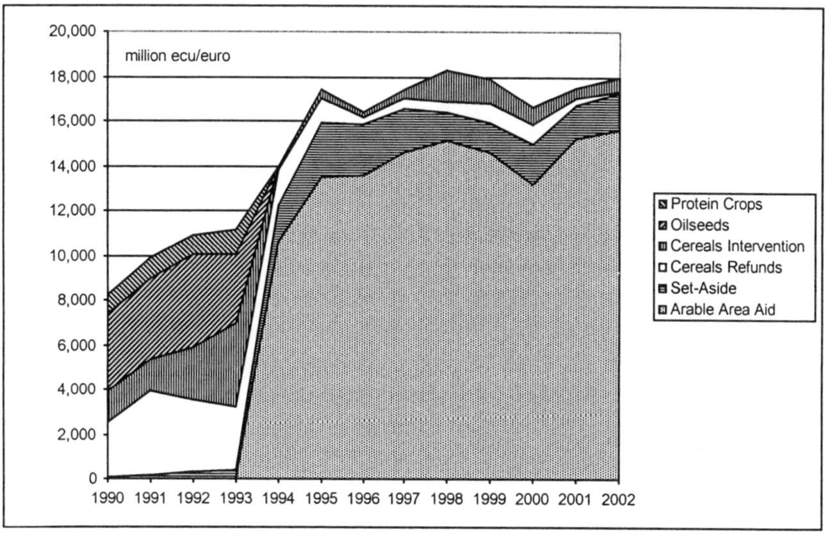

Fig. 2.1. EU budget expenditure on cereals, oilseeds and protein crops (million ecu/€, 1990-2002). Source: derived from Tables 3.4.3.1 and 3.4.4 of *The Agricultural Situation in the Community/European Union*, annual reports, and Annex 11 of the *Rapport financier concernant le Fonds européen d'orientation et de garantie agricole (F.E.O.G.A.) - Section Garantie*, annual. However reconciliation of data may be imprecise. 2002: budget rather than expenditure data

agreed in May 1992, the new system applied for crops grown and harvested in 1993, with payments to farmers in the latter part of 1993. Because of the complex accounting arrangements for the EU budget, the first charge on the budget appeared in 1994.

When Agra Europe (1991) first leaked news of MacSharry's plans in January 1991, it suggested that it was concern about the CAP's escalating budget costs, and not the international trade negotiations, that lay behind the initiative; indeed the work of Ross (1995) and Kay (1998), noted earlier, suggests that this had been a long-term concern of the Commission. However, the out-turn depicted in Fig. 2.1, and in particular the dramatic increase in expenditure between 1993 and 1995, clearly defied this aim. Why was this so?

In part the explanation lies in the fact that MacSharry's proposals were only partly implemented. The Commission had suggested that 80% of support was captured by 20% of recipients - a credible and widely accepted, but at the time undocumented, statistic - and that the reforms were in part designed to redress this imbalance (Commission, 1991a). The leaked text of January 1991 suggested that compensation would only be paid in full on the first 30 hectares of claims, at 75% on the next 50 hectares, and at 65% of the full rate on the remainder (Agra Europe, 1991). But in the Commission's formal proposals for CAP reform, issued later in the year, this provision had disappeared: there would be no size restriction. In its formal proposal, the Commission suggested that area payments on set-aside land would be restricted to 7.5 hectares; but this limitation was dropped by the Council when it adopted the package in May 1992. Furthermore, the area payment on set-aside was subsequently increased (see Swinbank and Tanner, 1996). These proposals, limiting payments to larger farmers, were referred to as *modulation* at the time: a word that acquired a rather different meaning in the debate over the 1999 reforms.

Had MacSharry's original ideas on modulation prevailed, some budget savings would have ensued; but a generous compensation package, switching the burden of support from consumers to taxpayers, was bound to result in an increase in taxpayer expenditure. Thus, although they marked a major shift in the CAP at the time, the introduction of arable area payments did create new problems, and leave MacSharry's successors with a difficult legacy. In particular, arable area payments:

- set a precedent for levels of farmer compensation that would be difficult to emulate in other sectors (sugar, etc.);
- absorbed a large swath of budget expenditure - about a third of all EAGGF guarantee expenditure - potentially crowding out other expenditure initiatives;
- by being based on past yields and cropping patterns, perpetuated the imbalance in the CAP under which 80% of the support is captured by 20% of producers;

- resulted in large budget transfers between Member States, such that countries with a net export surplus of cereals and oilseeds (and hence net recipients of budget transfers) were reluctant to change the policy;
- were depicted as a permanent feature of the CAP, in that the legislation they were based on was of unlimited duration and there was no provision for a reduction in payment levels (although many doubted the political sustainability of the system);
- kept land in arable crops, simply to 'farm' the subsidy, and became capitalised in land values and reflected in the levels of farm rents;
- raised expectations in the candidate states of Central and Eastern Europe that farmers there would benefit too; and
- raised problems in the WTO.

The bond scheme is designed to address these concerns, which are most acute with arable area payments. However, headage payments too present difficulties.

Headage payments

Although the 1992 reforms made important changes to the headage payment system, in fact these payments were in place before then. For sheepmeat, the ewe premium (a headage payment on each eligible female sheep) had been introduced in the 1980s, with payment limits from 1990. Until the creation of the Single Market, a different headage payment had applied for sheep in Great Britain (and similarly for bovines in the UK). From 1990 each farm received payment in full on the first 1,000 animals in Less Favoured Areas (LFAs), and 500 elsewhere, and at 50% of the full premium rate on additional animals. Furthermore, producers in LFAs received a Less Favoured Area supplement (the *rural world premium*) on their ewe premia up to the headage limit of 1,000 ewes. This was in addition to any hill livestock compensatory allowances Member States chose to make in LFAs, prior to the Agenda 2000 reforms. Premiums varied on a year-to-year basis, reflecting movements in the market price of sheepmeat and a fixed basic price. This policy proved popular with producers, and output expanded. In its first paper outlining its proposals for the MacSharry reforms, the Commission (1991a) noted that the budgetary cost of the policy had doubled over a four-year period. The rapid escalation in budget costs during the 1980s is evident in Fig. 2.2.

The main change introduced by the MacSharry reforms was that - within the limits set out above - an individual farmer's eligibility to payments became limited to the number of ewe premiums claimed in a reference year. Thus the right to receive premiums became limited by quota, which can be transferred between producers by sale (Commission, 1993). The regime was untouched by the Agenda 2000 reforms; but in December 2001, after a prolonged debate, the Council introduced a flat-rate payment to replace the variable premium that had previously applied (Agra Europe, 2001b).

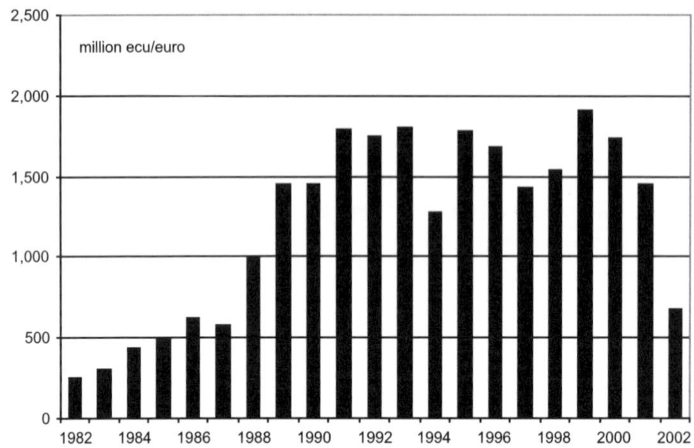

Fig. 2.2. EU budget expenditure on sheep and goat meat (million ecu/€, 1982-2002). Source: derived from Tables 3.4.3.1 and 3.4.4 of *The Agricultural Situation in the Community/European Union*, annual reports, and Annex 11 of the *Rapport financier concernant le Fonds européen d'orientation et de garantie agricole (F.E.O.G.A.) - Section Garantie*, annual. However reconciliation of data may be imprecise. 2002: budget rather than expenditure data

For beef producers the story is more complex. From 1980 a suckler cow premium had been in place under which producers who kept approved breeds for producing calves for beef production, and who did not milk their cows, could receive an annual subsidy. Later, in 1987, a beef special premium was introduced under which producers rearing steers (castrated males) or bulls for beef production could receive a subsidy (CAP Monitor, 1991).

In the 1992 reforms, following successive attempts in the 1980s to reduce the effective level of support offered by intervention, the Council cut the intervention price by 15% over three years from 1993. Beef premia were augmented to compensate for this, and various limitations introduced.

Thus the suckler cow premium was increased, but producer claims were limited by quota to the numbers claimed in a reference period (as with the ewe premium). The beef special premium was increased, but claims were limited to 90 animals per farm, and a count of the 'regional reference herd' for a base period was undertaken. Subsequently, if the total number of claims in the region exceeded this base, then claims were scaled back proportionally.

Furthermore, 'extensification' criteria were introduced for the beef premium. *First*, an individual farm's claim was restricted by stocking density limits. Initially set at 3.5 livestock units per forage area, this was reduced to 2.0

livestock units per forage hectare in 1996. In determining the number of livestock units, not only was a count taken of bovine animals on which beef premia were being claimed, but also the numbers of ewes under the sheep premium (at one ewe = 0.15 livestock units), and an allowance for the number of cows deemed to be required to produce the farm's milk quota (i.e. the farm's dairy quota divided by 5,050 litres). *Second*, if a farm's actual stocking density fell below 1.4 livestock units per forage hectare an *extensification* premium became payable in the form of a top-up to the suckler cow and beef special premia (Meat and Livestock Commission, 1993). There is also a *deseasonalisation* premium payable in those Member States where most animals tend to be slaughtered in the autumn, in order to encourage spring slaughter.

The 1999 (Agenda 2000) reforms saw a further (20%) reduction in the intervention price for beef, increases in the beef special and suckler cow premia, the introduction of a new slaughter premium, and a tightening of the criteria for the payment of the extensification premium (European Commission, 2001a). An innovation was that part of the compensation for the reduction in market price support was paid to Member States, a so-called *national envelope*, enabling the Member State to determine the criteria for making supplementary payments to producers under broadly defined EU rules. The UK has used this provision, for example, to relax the EU provision restricting claims under the beef special premium to 90 animals per farm.

Inevitably complex rules apply, under the IACS system, to ensure that farmers do have the livestock they claim, the appropriate entitlement to claim ewe and suckler cow premia, and an appropriate forage area (that is land not needed for the dairy herd to produce the farm's quota entitlement, or claiming arable area payments). Farmers rent in forage area, even if they do not intend to graze or harvest the grass, and engage in complex calculations to ensure that they maximise their total subsidy claim. Also, they must ensure that the appropriate number of animals are kept on their farm for the appropriate retention period. (MAFF, 2001 gives a graphic picture of the qualifying periods and other scheme dates.) Thus it has been suggested that the need to juggle livestock numbers might have been a contributory factor in the spread of foot and mouth disease in the UK in 2001.

Beef intervention and export refunds remain important policy instruments; nonetheless Fig. 2.3 clearly shows that headage payments have become the dominant cost of the beef regime.

A forage area payment?

Environmental (and animal welfare) groups have long expressed concern about headage payments: it is that, despite the extensification measures, over-stocking can result as farmers seek to maximise their subsidy claims, leading to over-grazing of sensitive habitats and malnutrition of stock (Baldock *et al.*, 2002). Thus, from time to time, it has been proposed that headage payments be converted into area payments: a forage area or grassland premium. This could either be determined on a flat rate basis per Member State (or geographical area within a Member State), or related to the agricultural capacity of the land. It would, however, be difficult to avoid creating gainers and losers: some farmers might receive more subsidy, and some less, with a scheme of this sort. Area payments would be more decoupled than headage payments, and consequently some farmers might choose to reduce their stocking density and thereby generate the desired environmental and animal welfare improvements.

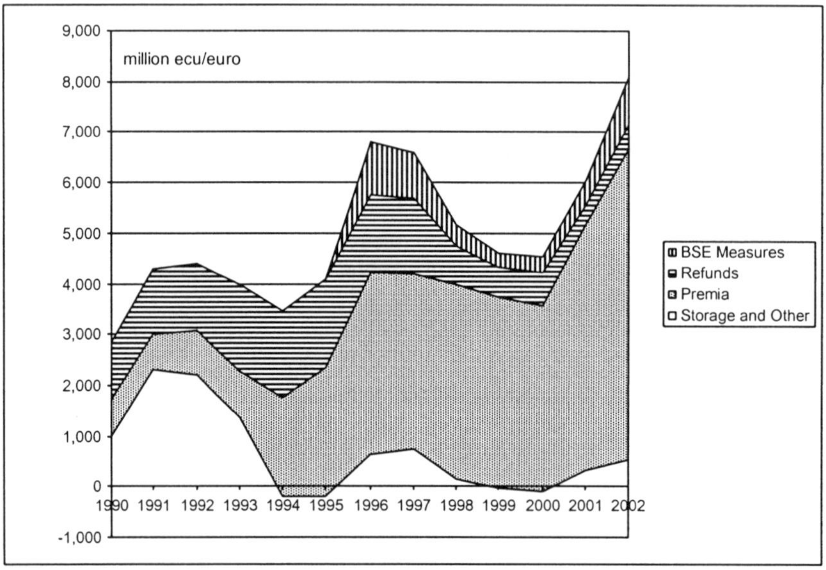

Fig. 2.3. EU budget expenditure on beef (million ecu/€, 1990-2002). Source: derived from Tables 3.4.3.1 and 3.4.4 of *The Agricultural Situation in the Community/European Union*, annual reports, and the *Rapport financier concernant le Fonds européen d'orientation et de garantie agricole (F.E.O.G.A.) - Section Garantie*, annual. However reconciliation of data may be imprecise. 2002: budget rather than expenditure data. 'Storage and Other' is negative in some years

Cross-compliance and modulation

One potentially important change introduced by Agenda 2000 was the requirement that 'with a view to better integrating the environment into the common market organisations, Member States should apply appropriate environmental measures in relation to agricultural land and agricultural production subject to direct payments'. Cross-compliance and modulation apply to all direct payments, including for example production aid for tobacco and seeds, and area aid for dried grapes and hops, as well as the arable area aids and beef and sheep premia referred to in this chapter (Council, 1999b). This could include 'specific environmental requirements constituting a condition for direct payments'. This provision, often dubbed 'cross-compliance', enabled Member States to 'decide on the penalties that are appropriate and proportionate to the seriousness of the ecological consequences of not observing the environmental requirements', which could extend to a complete forfeiture of the direct payment otherwise receivable by the farm (Council, 1999b). There is little evidence to suggest that Member States changed their implementation of area and headage payments to take advantage of this provision, and there is no data in the public domain to suggest that farmers had their direct payments reduced as a consequence. Indeed, Franz Fischler (2001) noted that only 'a few' Member States had implemented this provision, and commented 'Europe can and should go further'.

Modulation allowed Member States to reduce, by up to 20%, the direct payments that farmers would otherwise have received, where:

- the labour force used on their holdings during that calendar year, expressed in annual work units, falls short of limits to be determined by the Member States; and/or
- the overall prosperity of their holdings during that calendar year, expressed in the form of standard gross margin corresponding to the average situation of either a given region or a smaller geographic entity, rises above limits to be decided by Member States; and/or
- the total amounts of payments granted under support schemes in respect of a calendar year exceed limits to be decided by Member States (Council, 1999b).

Any expenditure savings that result from these actions are retained by the Member State for additional funding on Rural Development Programmes, the so-called Second Pillar (or Pillar 2) of the CAP, with the important proviso that the Member State must provide matched funding (i.e. 50% of budget cost is met by the Member State) (Harper Adams University and SAC, 2002).

In the UK, the system operates as a flat-rate abatement on all direct payments, reaching 4.5% in 2006. France had introduced a complex scheme, affecting about 10% of French farms, with an effective abatement rate of 5% averaged over those farms. However, following the re-election of Jacques Chirac in 2002, the incoming French farm minister announced a suspension of modulation in France until the outcome of the then pending mid-term review of Agenda 2000 had been settled (Agra Europe, 2002d). Portugal announced its intention of introducing modulation in 2003; and in Germany the Federal Government announced its intention of introducing modulation, but had difficulty reaching an accord with the Länder (Agra Europe, 2002a). (See also Harper Adams University and SAC, 2002.)

Small farmers' scheme

A potentially important change was introduced in 2001 as part of the drive for simplification of the CAP. This allowed Member States, for a trial period of three years, to make farm payments of up to €1,250 per year on the basis of one initial application from the farm. Farmers would not make *annual* IACS applications, only the initial claim based upon historical entitlements, and would not be required to grow crops or keep livestock to qualify. According to the Commission at the time, this facility could potentially benefit 20% of the EU's agricultural holdings and - by decoupling payments from production - move payments from the blue to the green box in the WTO (European Commission, 2001b). It is unclear which, if any, Member States have made use of this facility. However, data released in October 2002 suggests that 37% of recipients in 2000 received a payment of €1,250 or less (see Table 2.4 later in this chapter) (European Commission, 2002).

Direct payments dominate the CAP budget

As a result - primarily - of the 1992 and 1999 reforms, the pattern of spending on the CAP has changed radically: expenditure on export refunds and domestic market price support has been substantially displaced by direct payments, and there has been a marked switch from consumer to taxpayer support. Table 2.3 details budget expenditure on the CAP in 2000. Direct aids, largely made up of the headage payments and arable area payments discussed in this chapter, but also including olive oil, tobacco, and other subsidy programmes, account for 61% of the total. Together with price support measures (export subsidies, intervention buying, etc.), total spend on Pillar 1 support amounted to €36.3 billion. Pillar 2 expenditure, out of the Guarantee section of the EAGGF, under the Rural Development Regulation, accounted for just over 10% of EAGGF Guarantee expenditure; and small sums spent under the Guidance section of EAGGF made up the remainder.

Table 2.3. CAP budget expenditure, 2000

row:		€m	%
1	Price support	10,761.2	25.7
2	Direct Aids	25,529.2	61.0
3 = *1* + *2*	Pillar 1	36,290.4	86.7
4	Pillar 2	4,176.4	10.0
5 = *3* + *4*	EAGGF Guarantee	40,466.7	96.7
6	EAGGF Guidance	1,387.3	3.3
7 = *5* + *6*	Total	41,854.0	100.0

Source: Tables 3.4.1 and 3.4.2 from DG Agriculture's web site. Some rounding errors.

The pattern of expenditure on direct payments by Member State reflects in part the relative size of the agricultural sectors, but also the importance of cereals (and other arable crops) and yields. Thus France clearly emerges as the EU country that has the largest slice of EU funding from arable area and headage payments, as illustrated in Fig. 2.4. Also apparent is the extent to which a country such as Ireland is a major recipient of payments under the livestock schemes, but draws relatively little arable area aid.

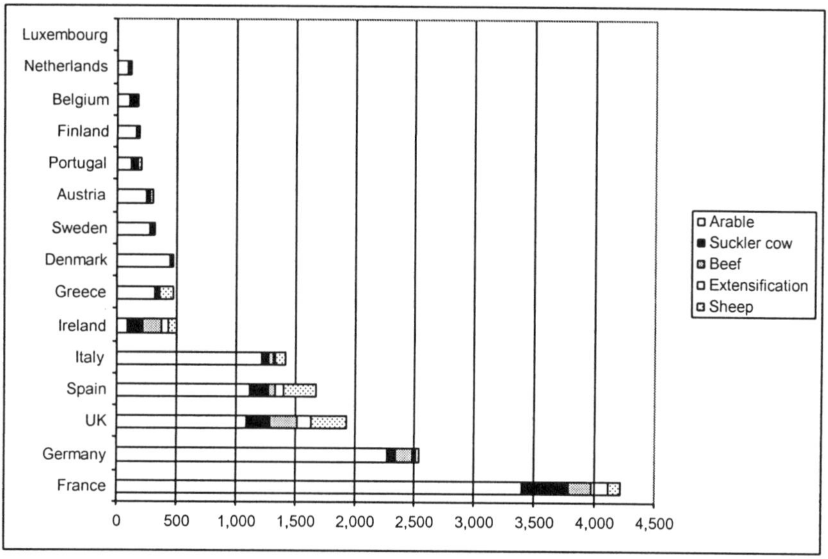

Fig. 2.4. Expenditure on direct payments, by member state (1999. £m). Source: MAFF (2001). £1 = €1.52 in 1999.

Figure 2.5 retains the same country order as Fig. 2.4, but records arable area and headage payments per hectare of utilised agricultural area. Although this denominator is far from ideal, it does attempt to correct for the relative overall size of the national agricultural sectors. Its effect is to equate France and Germany, and promote Denmark to the pole position as the main beneficiary under the scheme. Expressed in terms of utilised agricultural area, payments to Portugal amount to only 30% of those to Denmark. Thus it is no surprise that Portugal has expressed the view that it is a Member State 'severely handicapped by the imbalance in EAGGF Guarantee support' (as quoted in Agra Europe, 2002e).

The system of direct aids has perpetuated not only the distribution of support between Member States, but also that between individual claimants. In the US, a non-governmental organisation, the Environmental Working Group (EWG), has posted on its web site (www.ewg.org) details - *by recipient* - of all subsidy payments made between 1996 and 2000, revealing that 20% of recipients received 84% of all subsidies (Ayer and Swinbank, 2003). Until October 2002, data for the EU showing distribution of payments by size of payment was not readily available, and even then, in the new data base released, individual payments (and recipients) were not identified. However, periodically in the British House of Commons, Ministers have been asked to give details of payments by size of payment. Thus, for example, it was revealed that, in England in 1996, there were six payments of arable area aid of £1 million or more (House of Commons, 1997).

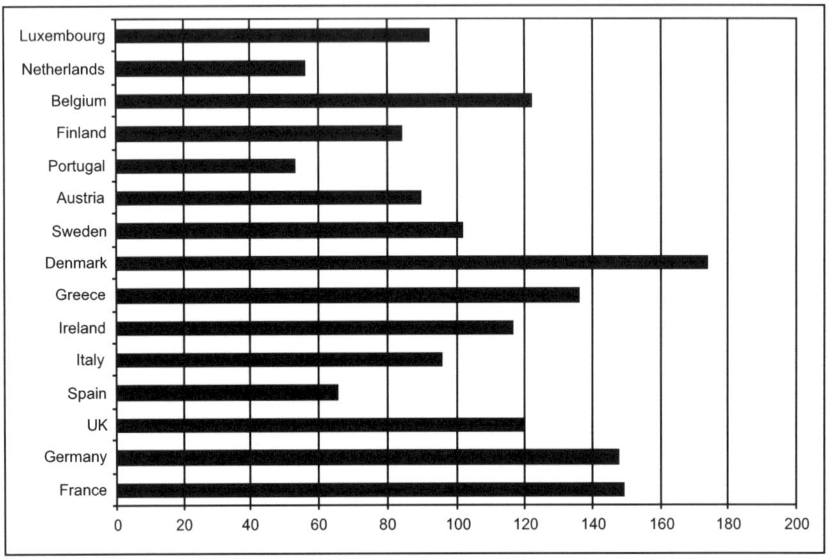

Fig. 2.5. Expenditure on arable area and headage payments, by member state (£/ha. 1999). Source: Data from Fig. 2.4, divided by 1998 utilised agricultural area

Commission data released to the European Parliament in October 2002 gives details of IACS payments for arable crops and livestock premia in the year 2000 (European Commission, 2002). The Commission cautions that, whilst the data is adequate for the purpose of financial control, not all Member States have yet deployed a 'system for unique identifiers', and thus the number of claimants may be over-stated. With this caveat in mind, Fig. 2.6 shows the different average payment levels by Member State: whilst the average payment per claimant in EU-15 amounted to €6,600, it ranged from €1,800 in Portugal to €18,900 in the UK.

Furthermore, the concentration of payments was particularly pronounced in Portugal and the UK. In Portugal, almost 80% of recipients collected €1,250 or less, accounting for less than 17% of payments. In the UK, 2.8% of recipients were paid €100,000 or more, and received nearly 25% of all the monies paid. The Commission's caution that the number of beneficiaries may not be accurately recorded, may be relevant in this comparison. Across the EU as a whole, 1.6% of recipients received nearly 24% of all payments, as revealed in Table 2.4.

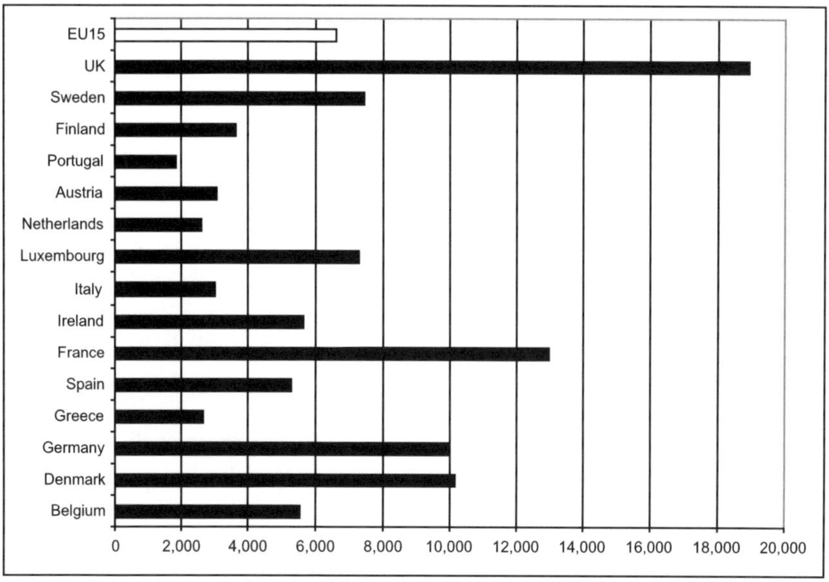

Fig. 2.6. Average IACS payments per claimant, €, 2000. Source: European Commission (2002)

Table 2.4. IACS payments categorised by size of payment, 2000

Payment Level: €	% of Claimants	% of Total Payments
> 0 and < 1,250	37.11	3.11
> 1,250 and < 2,000	10.84	2.62
> 2,000 and < 5,000	22.78	11.47
> 5,000 and < 10,000	12.60	13.66
> 10,000 and < 20,000	9.43	19.93
> 20,000 and < 50,000	5.59	25.40
> 50,000 and < 100,000	1.20	12.15
> 100,000 and < 200,000	0.28	5.65
> 200,000 and < 300,000	0.06	2.02
> 300,000 and < 500,000	0.03	1.97
> 500,000	0.02	2.03

Totals do not add to 100% because negative payments (i.e. repayments) reported in the original source have been ignored. The original > and < notation has been retained, and so it is not clear, for example, whether a payment of €1,250 would fall in the first, or the second, row. Source: European Commission (2002)

Milk

In February 1991, in its *Reflections* paper on what was to be the MacSharry Reforms, the Commission had merely stated that the market situation for milk meant that quotas would have to be reduced (Commission, 1991a). By the time the detailed proposals for reform had been tabled later in the year, however, the proposal contained two components. The Commission proposed that support prices be reduced by 10%, and quota allocations by 4%. In compensation for the price cuts, the Commission proposed an annual *dairy cow premium*. In compensation for the quota cut the Commission proposed that farmers should receive an annual compensation payment of 5 ecu/100 kg over a 10 year period. These arrangements would be operated through a bond issued to the farmers concerned, on the basis of which the Community would make annual payments over its life-time (10 years). The farmers could choose to keep the bond and receive the associated annual payments, or could sell it on the private market (Commission, 1991b).

The bond would have had all the characteristics of the bond scheme advocated in Chapter 4. In the event, the Council rejected both the proposal to reduce support prices, and that to reduce quota allocations, and so the Commission's proposed bond scheme for dairy farmers attracted little discussion in the Council. In Chapter 6 we discuss the fate of the Danish Government's proposed bond scheme for cereal farmers, advanced in 1991/92 as an alternative to arable area payments, but it remains unclear why the Commission favoured a bond scheme for dairy, but not for cereals, at the time.

In Agenda 2000 the Council agreed a 15% cut in the intervention prices for butter and skim milk powder, in three stages beginning in 2005. Farmers however were to be compensated for the loss in revenues, and from 2005 a new *dairy premium* was to be introduced, tied to a producer's quota holding (European Commission, 2001a). These decisions were revisited in the Fischler reforms of June 2003. The Commission has claimed that this new dairy premium will be a blue box payment, as 'The aid will be based on the individual quota rights in 1999/2000' (WTO, 1999).

Blue, green and amber boxes and the WTO

The *Agreement on Agriculture* concluded as part of the Uruguay Round GATT negotiations introduced limits on the total level of domestic support (an Aggregate Measurement of Support (AMS), the so-called amber box) that countries could bestow upon their farm sectors, whilst at the same time defining a set of criteria by which policies would be deemed to be exempt from the AMS limits (Josling *et al.*, 1996; Swinbank and Tanner, 1996). Thus, from 2000, the EU-15 has an AMS binding of €67.2 billion a year. Table 2.5 shows how the AMS constraint has reduced during the implementation period, and also documents the EU's AMS declarations (e.g. showing an AMS of €47.9 billion in 1999/2000).

Table 2.5. The EU's green, blue and amber box declarations

Million ecu/€	1995/96	1996/97	1997/98	1998/99	1999/2000
AMS Commitment	78,672.0	76,369.0	74,067.0	71,765.0	69,463.0
AMS Declared	50,026.0	51,009.0	50,194.0	46,683.0	47,885.7
Blue Box	20,845.5	21,520.8	20,442.8	20,503.5	19,792.1
Green Box	18,779.2	22,130.3	18,166.8	19,168.0	19,930.5
Document: G/AG/N/	EEC/12 /Rev.1	EEC/16 /Rev.1	EEC/26	EEC/30	EEC/38

Source: EU submissions to the WTO, as indicated by the document references in the bottom row

For the purposes of this study, the most important exemptions from the AMS calculations and constraints are the so-called green box measures listed in Annex 2 to the *Agreement on Agriculture*, and the so-called blue box measures defined in Article 6(5). (The *Agreement on Agriculture*, together with all other WTO Agreements, can be found on the WTO website at www.wto.org, as can the other WTO documents referred to in this chapter.) Annex 2, with its green box measures, was a central feature of the Dunkel text of December 1991; but paragraph 5 of Article 6 was a completely new insert by the Americans and the Europeans at Blair House (Swinbank and Tanner, 1996). Although the blue box provisions do not explicitly mention either the US deficiency payment system, or the EU's area and headage payments, it is widely recognised that the text was drafted to encompass both.

Article 6(5)(a) states:

'Direct payments under production-limiting programmes shall not be subject to the commitment to reduce domestic support if:

(i) such payments are based on fixed area and yields; or
(ii) such payments are made on 85% or less of the base level of production; or
(iii) livestock payments are made on a fixed number of head.'

The EU declares its arable area payments under (i), as detailed in Table 2.6, and its headage payments under (iii). Presumably it intended to declare its new dairy premium under (i), although the wording of Article 6(5) does not quite suit. Option (ii) was written to reflect the deficiency payments policy then in operation in the US.

Clearly, if annual blue box expenditure of €19.8 billion had to be added to existing AMS support of €47.9 billion per annum, the EU's AMS constraints would become binding (and even more so following EU enlargement). Thus the

EU had a clear interest in maintaining the blue box (or refashioning its blue box payments so that they fell within the green box).

Table 2.6. The EU's blue box expenditure, as notified to the WTO

Million ecu/€	1995/96	1996/97	1997/98	1998/99	1999/2000
Arable Area Payments					
Maize	973.0	1,222.8	1,212.7	1,182.2	1,159.0
Other cereals	8,638.6	10,001.2	9,554.7	9,372.1	8,841.6
Oil seeds	2,381.0	2,439.4	2,368.6	2,263.7	1,318.0
Protein crops	522.7	525.0	617.8	647.2	524.3
Non-textile flax	72.4	96.5	129.2	165.5	306.6
Set-aside	2,112.1	1,827.8	1,251.3	1,272.3	1,848.0
Durum wheat supplement	948.3	1,080.6	1,016.2	993.2	1,006.2
Rice			40.5	81.3	124.3
	15,648.1	17,193.3	16,191.0	15,977.5	15,128.0
Livestock Payments					
Suckler cows	2,446.4	2,042.9	1,694.9	1,669.4	1,628.4
Beef premium	1,407.2	1,238.5	1,340.8	1,297.3	1,299.3
Deseasonalisation	23.0	39.5	45.1	23.7	2.6
Ewes and goats	1,320.8	1,006.6	1,171.0	1,535.6	1,733.8
	5,197.4	4,327.5	4,251.8	4,526.0	4,664.1
TOTAL BLUE BOX	20,845.5	21,520.8	20,442.8	20,503.5	19,792.1

Source: as Table 2.5

Critics claim that blue box policies are not completely decoupled - the crop has to be grown and the animals kept if payments are to be claimed - and therefore the exemption should be seen as a temporary expedient of the Uruguay Round which should be phased out in the Doha Development Agenda[1]. In practice the blue box provisions could only be removed from the *Agreement on Agriculture* with the EU's agreement: the WTO proceeds on the basis of consensus, and the existing blue box measures are not time limited.

The Peace Clause (Article 13 of the Agreement, headed 'Due Restraint'), however, is time limited. It expired at the end of 2003 (or the relevant 2003/04 marketing year). The Peace Clause is complex, untested in Dispute Settlement proceedings, and difficult to understand, but basically it provided a measure of protection to green box policies (section (a) of Article 13), AMS and blue box measures (section (b)) and export subsidies (section (c)) where there might otherwise be a conflict between the provisions of the *Agreement on Agriculture* and other GATT/WTO rules.

Thus, whilst the Peace Clause remained valid, blue box measures were not actionable subsidies under the Subsidies Agreement, and could not lead to actions based on the nullification or impairment of tariff concessions enjoyed by other WTO Members, *provided* 'such measures do not grant support to a specific commodity in excess of that decided during the 1992 marketing year'. Without the Peace Clause it might be claimed that blue box payments circumvent the export subsidy constraints on agricultural products (the policy does not differentiate between products consumed within the domestic market and those exported, and Australian authors have often referred to them as implicit export subsidies (e.g. Roberts *et al.*, 1999)), or that, in promoting domestic production, they have infringed the tariff concessions of another WTO Member. It will be recalled from the earlier discussion in this chapter that at Blair House the EU in effect conceded that its area payment scheme for oilseeds impaired tariff concessions enjoyed by the US, and sought a *rapprochement* with the US. A replay might involve any product covered by blue box payments, and any country with a substantial export interest.

Even if the Peace Clause were rolled over in its entirety, it was debatable whether its provisions would have extended to blue box payments in the new Central and Eastern European Member States following EU enlargement. Would the condition 'such measures do not grant support to a specific commodity in excess of that decided during the 1992 marketing year' be respected in the sense that the unit rate of EU subsidy (regardless of the size of the EU) remained unchanged, or would it be infringed because blue box payments would now be paid to a large number of additional farmers who had not benefited from this level of support in 1992?

The US, together with the EU, the joint architect of the blue box, no longer makes use of the blue box provisions, and has proposed its abolition through merger with the AMS (WTO, 2000a). The EU is almost alone in making use of the provision: according to the WTO only Iceland, Norway, Japan, the Slovak Republic and Slovenia invoke the blue box in making their submissions (WTO, 2002), and the latter two are scheduled to become EU members in 2004. Although an extension of the Peace Clause for green box measures can be envisaged, a parallel extension for blue box measures and export subsidies is less easy to see. In its *Comprehensive Negotiating Proposal* of December 2000, the EU had proposed that the concept of the blue box should be maintained (WTO, 2000b). However, as a result of its mid-term review of Agenda 2000, in the Fischler reforms of June 2003, the EU has largely abandoned its use of blue box measures.

The green box

Whilst blue box payments are only partially decoupled, the overarching requirement (as specified in paragraph 1 of Annex 2) is that green box measures have 'no, or at most minimal, trade-distorting effects or effects on production'. In practice, giving operational significance to the term 'minimal trade-distorting effects' will not be easy.

Annex 2 sets out a whole series of policy measures that might qualify under this umbrella provision, together with further detailed criteria that must be met in addition to the generic requirement of paragraph 1. Paragraphs 2 to 4 cover 'General services' (e.g. research and extension services), 'Public stockholding for food security purposes', and 'Domestic food aid'. Paragraphs 5 to 13 deal with payments to individual producers, for example 'Structural adjustment assistance provided through producer retirement programmes' (paragraph 9).

Paragraph 5 reads:

'Support provided through direct payments (or revenue foregone, including payments in kind) to producers for which exemption from reduction commitments is claimed shall meet the basic criteria set out in paragraph 1 above, plus specific criteria applying to individual types of direct payment as set out in paragraphs 6 through 13 below. Where exemption from reduction is claimed for any existing or new type of direct payment other than those specified in paragraphs 6 through 13, it shall conform to criteria (b) through (e) in paragraph 6, in addition to the general criteria set out in paragraph 1.'

In practice, the word 'decoupled' is used only once, in the heading 'Decoupled income support' (paragraph 6). It is under this heading that a bond scheme would be declared. Decoupled income support payments would have to meet the following criteria:

(a) Eligibility for such payments shall be determined by clearly defined criteria such as income, status as a producer or landowner, factor use or production level in a defined and fixed base period.
(b) The amount of such payments in any given year shall not be related to, or based on, the type or volume of production (including livestock units) undertaken by the producer in any year after the base period.
(c) The amount of such payments in any given year shall not be related to, or based on, the prices, domestic or international, applying to any production undertaken in any year after the base period.
(d) The amount of such payments in any given year shall not be related to, or based on, the factors of production employed in any year after the base period.
(e) No production shall be required in order to receive such payments.

We believe that the bond scheme outlined in Chapter 4 would have no difficulty meeting these provisions. Nonetheless, the language used in Annex 2 clearly indicates that the drafters of this provision had not seen an outline of a bond scheme. For example, Paragraph 5 is headed 'Direct Payments to *Producers*', and Paragraph 6 'Decoupled *Income* Support' (emphasis added), neither of which encapsulates the idea of making annual payments to bond holders whose links with agriculture may be distant or non-existent. Paragraphs 5 and 6 do not specifically recognise that farms might be limited liability companies, rather than in single ownership or in partnerships. However, we very much doubt that these semantic differences would lead one of the EU's trading partners to challenge the validity of a bond scheme, or, if challenged, for such a challenge to succeed.

There are, nonetheless, those who believe that the provisions are too broadly drawn. India, for example, has proposed a revision of the *Agreement on Agriculture* so that direct payments under paragraphs 5 and 6 of the green box, as well as blue box payments, would be included in the AMS (WTO, 2001a). This would be a more dramatic re-drafting of the green box than envisaged by most WTO Members, and in our judgement it is unlikely to prevail. To many developing countries, however, it is self evident that the sheer size of EU expenditure on green box policies (some €19.9 billion in 1999/2000 according to Table 2.5) inevitably distorts competition.

Reconciling budget expenditure, WTO declarations and OECD Producer Support Estimates (PSEs)

In addition to the budget costs of the CAP reported in Table 2.3, and the EU's declarations of green, blue and amber box support, the OECD on an annual basis produces a *Producer Support Estimate* (PSE). The PSE is an estimate of the extent to which farmers' revenues have been increased by domestic farm policies.

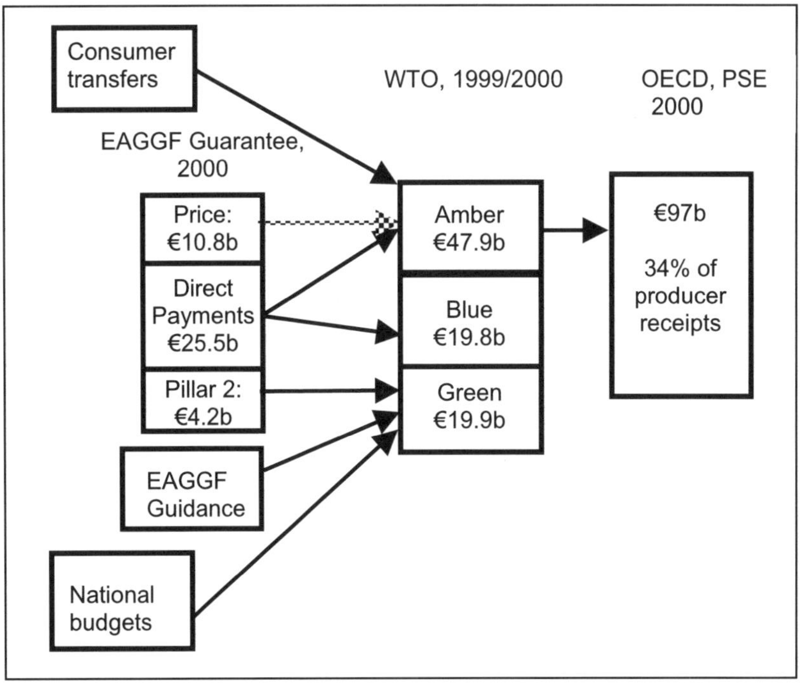

Fig. 2.7. Schematic reconciliation of CAP support, 2000. Source: Tables 2.3 and 2.5, and OECD (2002)

Reconciling these three sets of numbers is not always easy, and it is certainly not an exact science, but Fig. 2.7 attempts to demonstrate how the three sets of numbers relate to each other.

Some of the numbers reported are accounting data, taken from the reported expenditures of the EU and the Member States, whilst other numbers are estimates. Thus EU Budget expenditure under the EAGGF Guarantee and Guidance sections, for example, and blue and green box expenditures, fall into the former category, whilst the amber box and PSE try to capture the effect of all support, including the transfers from consumers (or more generally the users of agricultural raw materials). In compiling these estimates, however, the amber box uses a fixed world market price for reference purposes, whereas the PSE does not. Thus an increase in the world market price (in terms of euro) will not impact directly on the amber box calculation, but it would result in a fall in the PSE.

Not all direct payments fit the blue box: some are declared in the amber box. The green box embraces Pillar 2 and some Guidance section expenditure, and spending by Member States.

Eastern enlargement and the mid-term review

When the Council of Agriculture Ministers concluded its provisional agreement on the Agenda 2000 package on 11 March 1999, it undertook to conduct a mid-term review of the milk quota regime in 2003, and asked the Commission to submit a report of the operations of the policy on oilseeds by 2002 (Council, 1999a). In reducing the cut in the intervention price for cereals from 20 to 15%, the European Council in Berlin noted that 'A decision upon a final reduction in the intervention price to be applied from 2002/03 onwards will be taken in the light of market developments'; and it also invited the Commission to submit a report on agricultural expenditure in 2002, 'accompanied, if necessary, by appropriate proposals' (European Council, 1999). Thus, Franz Fischler, the Agriculture Commissioner, was given an opportunity to revisit CAP reform before the end of the Agenda 2000 implementation period (2000-2006).

By November 2000 it seemed that the Commission planned to concertina all the mid-term reviews into a single package, and that this would amount to more than a mere assessment. The French President, Jacques Chirac, who had almost single-handedly engineered the outcome of the Berlin European Council (Schwaag Serger, 2001), and Franz Fischler, clashed over the likely scope of the mid-term review, with Chirac adamant that further CAP reform could not be contemplated before 2006, and Fischler insisting that he had a mandate from the European Council to table 'appropriate proposals' (Agra Europe, 2000).

There were a number of ideas put forward for change. For example, in May 2001 the then Portuguese Agriculture Minister, Antonio Capoulas, advocated a 'New direction for European Agriculture' in which existing direct payments would, until 2006, continue to be paid as 'transitional income-guarantee aids' completely decoupled from production (Portuguese Government, 2001)[2]. From 2011 onwards, under the Portuguese proposal, only two strands of support would be available: (i) a new aid payment per farm, decoupled from production but 'oriented towards environment, employment and quality' and (ii) a farm income stabilisation scheme.

But with Presidential and Parliamentary elections due in France in 2002, it was un-politic for the Commission to launch a major debate on the future of the CAP until the French electoral cycle had been concluded. Thus, as the months slipped by, Fischler dampened down suggestions that the mid-term review would be more than a package of minor amendments. In March 2002, with the French Presidential election imminent, Fischler's *chef de cabinet* was suggesting that adjustment, rather than reform, would be the characteristic of the mid-term review (or MTR as it was increasingly dubbed) (Agra Europe, 2002b). The package of proposals that the College of Commissioners adopted on 10 July 2002 went well beyond that (Commission, 2002b).

The document contained some commodity specific suggestions, including a further cut in the intervention price for cereals, and a substantial cut in that for rice, with corresponding increases in direct payments. But the dramatic centre-piece was for a further decoupling of most direct payments (Commission, 2002b). This key feature of the MTR was retained in the formal proposals for CAP reform, submitted in January 2003 (Commission, 2003), and adopted by the Council in June 2003 (Council, 2003). In Chapter 9 we outline the progress of this reform - now known as the *Fischler Reform* - through the Council of Ministers, and here we briefly outline the salient features of the Council's decisions of June 2003. The new *Single Payment Scheme* is to be introduced from 2005, although Member States can defer application until 2007 (Council, 2003). However, Member States will have considerable discretion in applying the scheme.

The *basic* plan follows the Commission's original proposal, with entitlement based on IACS claims in the period 2000-2002 (enhanced by the new dairy premium). This entitlement will be attached to the farm, and - if the farm is subsequently split - so too will be the payment. Indeed, under this scheme, entitlement to the Single Payment Scheme will be allocated on a per hectare basis. The total IACS claim of the farm will be divided by the area of land that underpins that claim to give an entitlement of €x per hectare on y entitlement hectares. Annual claims have to be submitted in order to claim under the Single Payment Scheme; with entitlement hectares matched by 'farmed' hectares. Single Payment Scheme entitlements can be sold with or without land, but they are valueless unless the owner has eligible farmland to match against annual claims. The set-aside obligation of the old IACS scheme is carried forward into the new scheme (and applied on a pro-rata basis under the regionalised scheme outlined in the following paragraph), and farmers are not allowed to claim payments on land under permanent crops or fruit and vegetables, but otherwise they are free to farm. However, a basic requirement is 'that all agricultural land, especially land which is no longer used for production purposes, is maintained in good agricultural and environmental condition' (Council, 2003), and an enhanced, and mandatory, system of cross-compliance will apply. Some crop specific payments are retained.

Member States can opt to amend the basic scheme in two ways. First, rather than treat each farm as a single entity, with its own unique Single Payment Scheme entitlement, all of the Single Payment Scheme entitlements in a particular region can be pooled and then paid on a flat-rate per hectare basis on all eligible farmland in that region. Furthermore, arable area payments, livestock payments, and the new dairy premium, can each be paid out on either an historic entitlement or a regionalised basis.

Some Member States were concerned that, under 'full' decoupling, land might be abandoned in some regions, threatening the multifunctionality of agriculture. Consequently Member States can, if they choose, opt for 'partial implementation' under which part of the arable area and headage payments can be retained, under the old IACS scheme, with the remainder paid as a Single Payment Scheme. The options are complex, but up to 25% of the old arable area payment can be retained, for example. As yet it is unclear which Member States will make use of this option.

Under the original proposal, direct payments would have been subject to *dynamic modulation*. For any farmer the first €5,000 per annum would have been exempt, but on sums in excess of this modulation would apply: at 3% in the first year, rising to 20% in year seven as a result of annual increments of three percentage points (Commission, 2002b). In 2000, more than 70% of IACS claimants were paid €5,000 or less, with the proportion rising above 90% in Portugal (European Commission, 2002). In addition, the new Single Payment Scheme would have been capped. After deducting the €5,000 franchise mentioned above, the maximum payment per claimant would have been €300,000 (Commission, 2002b). In 2000, 0.05% of claimants had IACS claims in excess of €300,000, mostly concentrated in Germany but some in other Member States notably Spain and the UK, accounting for 4% of payments (European Commission, 2002). Monies clawed back in this way would be retained by the Member State, for use on Pillar 2 activities.

The agreed modulation package is more modest (Council, 2003). Modulation will, in 2007, amount to 5% of payments in excess of €5,000 per claimant. However, in addition, Member States may choose to reduce direct payments by a further 10% and redistribute the proceeds 'for specific types of farming ... important for the protection or enhancement of the environment or for improving the quality and marketing of agricultural products' (Council, 2003). Furthermore, a new *Financial Discipline* is to apply. If expenditure on Pillar 1 is forecast to exceed annual expenditure limits, less €300m, then the Council, on a Commission proposal, will be expected to 'adjust' the direct payments accordingly, so as to fit within the expenditure limits (Council, 2003). Whether this will happen remains to be seen.

Enlargement

At Copenhagen in December 2002 the EU concluded its negotiations with 10 applicant states, paving the way to enlarge the membership of the EU to 25 on 1 May 2004. The new Member States are: Cyprus, the Czech Republic, Estonia, Hungary, Latvia, Lithuania, Malta, Poland, Slovakia and Slovenia.

Although there were many questions that exercised the negotiators, whether or not farmers in the new Member States should qualify for direct payments was a key issue (Buckwell and Tangermann, 1999; Daugbjerg and Swinbank, 2004). There was a question of principle, and several of practicalities.

The question of principle went to the core of the CAP. Were the area and headage payments that pre-empted such an important part of the CAP's budget there to *compensate* for policy changes in 1992 (and 1999), or did they offer *income* support, a permanent feature of the CAP, contributing to the CAP's multifunctional role? If the former, then the case was weak for extending these payments to farmers who had not experienced the policy change: but this then raised a subsidiary question as to whether two quite different regimes could coexist in an enlarged EU? Would competition not be distorted if farmers in western Europe continued to receive compensation payments, whereas those in the east did not? And this subsidiary question was even more forcefully put if it was conceded that direct payments formed a permanent feature of CAP support. In its *Issues* paper of January 2002, which debated these concerns, the Commission (2002a) admitted that direct payments had 'lost part of their compensatory character after 10 years of implementation' and had 'instead become direct income payments'.

Additionally it was questioned whether or not the EU's budget could afford direct payments in the new Member States, and whether they would be compatible with the EU's obligations under the WTO *Agreement on Agriculture*. Mr Fischler suggested that in order to apply the CAP's supply control mechanisms in the new Member States, farmers there would have to be in receipt of direct payments: a set-aside obligation can only be imposed on a farm if that business is in receipt of area payments. Further concerns focussed on the economic impact of direct payments. If extended to the new Member States would this not create new inequalities between farmers (cereal growers in receipt of payments, and potato producers who were not, for example) and between the farm sector and other sectors of the economy? Would the sudden windfall of additional CAP subsidies help or hinder the very necessary farm restructuring that was needed in the new Member States? Furthermore, it was argued that direct payments would be capitalised into land values, and hence into the cost structure of farms in the new Member States as it had in EU-15. This would tend to 'lock-in' CAP support in the new Member States, as it had in EU-15, making future reform of the CAP more difficult, as investment decisions would have been taken on the expectation of a continuation of support (Daugbjerg and Swinbank, 2004). Indeed to some extent this had already happened, as a number of applicant states had increased their levels of farm support to align with the existing CAP.

In the event, and after a long prevarication, in January 2002 the Commission (2002a) proposed that direct payments should be extended to the new Member States, but in a phased manner starting at 25% of the payment level in EU-15 in 2004, and rising to the full EU-15 payment level after 10 years. Many of the applicant states were unhappy about this low percentage. Some existing Member States (the UK, Germany, Sweden and The Netherlands) suggested it would be more prudent to await the discussion on the mid-term review, giving a clearer steer about the future of direct payments, than to make a commitment at an early stage to the applicant states. But in the end the French position prevailed.

At the meeting of the European Council in October 2002, which confirmed the expectation that negotiations would be concluded with the ten at the Copenhagen meeting of the European Council in December 2002, it was decided that the EU would endorse the Commission's suggestion for a phased introduction of direct payments (European Council, 2002). The European Council also decided to impose tight financial limits on the CAP budget spend.

First, it reiterated the Agenda 2000 decisions, taken in Berlin in 1999, to place limits on Pillar 1 expenditure under the CAP through to 2006 (European Council, 2002). This has become known as category 1a expenditure, and specifically excludes rural development and accompanying measures. The phasing in of direct aids will not cause too many problems with this constraint, as the first payments will not be made until the 2005 budget year. However, beyond 2006, expenditure will mount (see, for example, estimates by Ackrill, 2003).

The second constraint agreed by the European Council is even more problematic. It is that category 1a expenditure for an EU of 25 Member States cannot exceed, in nominal terms, the Agenda 2000 budget limits for 2006 plus 1% per year in the period from 2007 to 2013 (European Council, 2002). This leaves scope for expanding the limits should any one of Bulgaria, Romania or Turkey gain EU membership in that time frame (and indeed the European Council could at some future time reverse its decision), but no clear guide as to how direct payments for 25 Member States are to be funded.

It is reported that these decisions led the French President to suggest that the Commission's proposals for the mid-term review should be set aside until 2006 (e.g. Financial Times, 2002). However, this was not agreed by the European Council, and - as already noted - in January 2003 the Commission tabled its revised proposals for the mid-term review, with 'dynamic modulation' replaced by 'degression'.

Thus enlargement was agreed (at Copenhagen in December 2002) with the CAP intact, although it had been the conventional wisdom of the mid-1990s that CAP reform would have to precede enlargement (see, for example, the Commission's *Agricultural Strategy* paper of December 1995). As proposed in the Commission's January 2002 *Issues* paper, farmers in the new Member States will receive direct payments on a phased basis, starting at 25% of the EU-15 rate in 2004 and rising to 100% of the then applicable EU-15 level in 2013, with some top-ups from national funds permitted.

However, the full rigours of the IACS system would not necessarily apply in all the new Member States, for the Copenhagen package endorsed a further element of the Commission's January 2002 proposal. It is that a 'simplified' scheme can be deployed by the new Member States for a period of three years (extendable to five), at their discretion. Under this scheme, a budget has been agreed for the new Member States based upon base areas, historic yields, and eligible cattle and sheep numbers, and then this sum can be disbursed on a flat-rate, decoupled basis, over the entire agricultural area (Commission, 2002a). This will be more administratively feasible than the highly bureaucratic IACS scheme, and would have given farmers in the new Member States that chose to implement this provision greater freedom in their farming operations than that enjoyed by farmers in EU-15 if the Commission's mid-term review proposal for a partially decoupled Single Payment Scheme had not been accepted.

Similar developments in North America

It is not just in the EU that a partial decoupling of support policies has been pursued in the 1990s, for in North America, both Mexico and the US introduced decoupled payments for farmers.

In Mexico, policy reform was prompted by entry into NAFTA (the North American Free Trade Area), and the prospect of a fall in farm-gate prices as a result of the consequently cheaper imports from the US and Canada[3]. From 1994, a programme of direct assistance for agriculture, PROCAMPO (Programa de Apoyos Directos al Campo), was applied. The purpose was to make NAFTA more attractive to farmers, and to provide them with adjustment assistance. All farmers who had planted any of nine basic crops (including maize, wheat and cotton) in the any of the three preceding years were entitled to claim PROCAMPO payments. Claims are re-submitted each year (indeed twice a year when two crops are grown within the year) for a flat-rate area payment on the basis of the area grown the previous year. Thus payments are decoupled from yields, and from areas grown in the current year, but the land must still be devoted to agricultural production, forestry or an approved environmental programme. Only farm households growing one of the nine basic crops in the period 1991-1993 are entitled to apply, and the maximum area payment they can claim is the area registered in 1994. Nonetheless, payments are made to approximately three million producers a year, and this has been a major cash transfer to the rural economy: even self-sufficient farms qualify for PROCAMPO payments, even though - with no marketable surplus - they had not benefited from the price support policies that preceded PROCAMPO. The original intent was that the *real* level of support would remain constant for ten years, and then be phased out over the following five years, allowing a fifteen-year transition. In practice, payment rates have not been sufficiently increased to reflect inflation. PROCAMPO qualification certificates are used as collateral for borrowings from banks, or input suppliers.

In the US, a new Farm Bill was agreed in 1996: the *Federal Agricultural Improvement and Reform* (FAIR) *Act*. This decoupled payments for the main arable crops, abolishing deficiency payments and replacing them with *Production Flexibility Contract* (PFC, also called Agricultural Market Transition Assistance (AMTA)) payments for farmers for the period 1996-2002. Individual farmers entered into a contract with the government to receive PFC payments based on 85% of their 1996 base areas of wheat, feed grains, rice and cotton, and previous farm bill yields for 1995. Payments were to peak in 1998, and thereafter decline; but they would remain significant even in the last year of FAIR (Roberts *et al.*, 1999). PFC payments no longer fell within the blue box (as their predecessor deficiency payments did), but instead into the green box.

However, with depressed commodity prices, in 1998 (and again in 1999, 2000 and 2001) the US made emergency payments to farmers that were clearly not fully decoupled. These *Market Loss Assistance Payments* were partially decoupled, because they were based on the same historic areas and yields used to compute PFC payments, but they were triggered by price movements. The Clinton Administration, apparently, could not decide how these 1998 payments should be declared; but in June 2001 the Bush Administration did admit they were amber box payments (Agra Europe, 2001a).

Although the 2002 Farm Bill, the *Farm Security and Rural Investment Act*, continues with decoupled direct payments (again based on 85% of producers' base areas), it does retain the loan rate programme and - for wheat, maize, grain sorghum, barley, oats, rice, upland cotton, oilseeds, and peanuts - it reintroduces a counter-cyclical subsidy abandoned by FAIR: if market prices fall below a set target price a deficiency payment is paid. However, in contrast to past deficiency payment programmes the new counter-cyclical programme decouples payments from production. Payments will be based on 85% of a fixed base acreage and a fixed historical yield (Ayer and Swinbank, 2002). With the benefit of hindsight it is difficult to endorse the view of Guyomard *et al.* (2000) that the FAIR Act represented a 'watershed' that would 'force the European Union to reform its agricultural policy so that compensatory payments ... are included in the green box'. Other pressures have brought about that change.

Notes

[1] Article 20 of the *Agreement on Agriculture* mandated new negotiations to continue the reform process, and these have been under way in Geneva since March 2000. The WTO's Fourth Ministerial Meeting in Doha launched a new round of multilateral trade negotiations (dubbed the Doha Development *Agenda* rather than *Round*) which has subsumed the Article 20 negotiations (Laird, 2002). It is supposed to be a *single undertaking* (that is all countries will be expected to

accept all aspects of the final package), which relieves an EU fear that it would be unable to secure trade-offs between sectors if the agriculture negotiations proceeded in isolation. But agriculture might yet be detached from the rest. A tight timetable had been set: for agriculture the 'modalities' (e.g. the rules setting out tariff, AMS and export subsidy reductions) were to be determined by 31 March 2003; and when WTO Ministers next met in Cancún, in Mexico, in September 2003 they were supposed to have lodged their detailed schedules of tariff and subsidy reduction commitments (WTO, 2001b). The failure to agree on the modalities by 31 March 2003 put this process in doubt (Josling, 2003). The whole is supposed to be concluded by 31 December 2004, which is an ambitious target given that 140+ WTO Members have to agree the package by consensus.

[2] But Ministers come and go. Early in 2002 the Portuguese Government changed following parliamentary elections, and the new Minister of Agriculture - Armando José Cordeiro Sevenate Pinto - proved to be much more defensive of the existing CAP. He joined with the French Minister (Hervé Gaymard) in writing to leading European newspapers in support of the CAP (Boden *et al.*, 2002).

[3] This paragraph draws heavily on Sadoulet *et al.* (2001). Mexico declares the PROCAMPO payments as decoupled income supports to agriculture, under paragraph 6 of the green box provision listed in Annex 2 to the *Agreement on Agriculture* (WTO, 2000c).

References

Ackrill, R.W. (2000) CAP reform 1999: a crisis in the making? *Journal of Common Market Studies* 38, 343-353.

Ackrill, R.W. (2003) EU enlargement, the CAP and the cost of direct payments: a note. *Journal of Agricultural Economics* 54, 73-78.

Agra Europe (1990) 12 October, E/4.

Agra Europe (1991) 11 January, P2.

Agra Europe (2000) 17 November, EP/7-8.

Agra Europe (2001a) 29 June, EP/6.

Agra Europe (2001b) 21 December, EP/1.

Agra Europe (2002a) 8 February, N/2.

Agra Europe (2002b) 22 March, EP/6.

Agra Europe (2002c) 10 May, EP/3.

Agra Europe (2002d) 24 May, N/1.

Agra Europe (2002e) 28 June, EP/5.

Ayer, H. and Swinbank, A. (2002) The 2002 U.S. Farm Bill: what's in it for CAP and WTO reform? *EuroChoices* 1, 26-32.

Baldock, D., Dwyer, J. with Sumpsi Vinas, J.M. (2002) *Environmental Integration and the CAP. A Report to the European Commission, DG Agriculture*. Institute for European Environmental Policy, London.

Boden, F., Canete, M.A., Pinto, A.J.C.S., Gaymard, H., Happart, J., Molterer, W. and Walsh, J. (2002) CAP is something we can be proud of. Letter to the Editor. *Financial Times*, 23 September 2002, 22.

Buckwell, A. and Tangermann, S. (1999) Agricultural Policy Issues of European Integration: the Future of Direct Payments in the Context of Eastern Enlargement and the WTO. *MOCT-MOST: Economic Policy in Transitional Economies* 9, 229-254.

CAP Monitor (1991) Issue of 8 October 1991. Agra Europe (London) Ltd.

Commission of the European Communities (1991a) *The Development and Future of the CAP. Reflections Paper of the Commission*. COM(91)100. CEC, Brussels.

Commission of the European Communities (1991b) *The Development and Future of the Common Agricultural Policy. Follow-up to the Reflections Paper, COM(91)100 of 1 February 1991*. COM(91)258. CEC, Brussels.

Commission of the European Communities (1993) *The Agricultural Situation in the Community. 1992 Report*. Office for Official Publications of the European Communities, Luxembourg.

Commission of the European Communities (1995) *Study on Alternative Strategies for the Development of Relations in the Field of Agriculture between the EU and the Associated Countries with a View to Future Accession of these Countries*. COM(95)607. CEC, Brussels.

Commission of the European Communities (2002a) *Enlargement and Agriculture: Successfully Integrating the New Member States into the CAP. Issues Paper*. SEC(2002)95. CEC, Brussels.

Commission of the European Communities (2002b) *Communication From the Commission to the Council and the European Parliament. Mid-term Review of the Common Agricultural Policy*. COM(2002)394. CEC, Brussels.

Commission of the European Communities (2003) *Proposal for a Council Regulation Establishing Common Rules for Direct Support Under the Common Agricultural Policy and Support Schemes for Producers of Certain Crops*. COM(2003)23. CEC, Brussels.

Council of the European Communities (1992) Council Regulation (EEC) No 1765/92 of 30 June 1992 establishing a support system for producers of certain arable crops. *Official Journal of the European Communities*, L 181.

Council of the European Communities (1999a) *2162nd Council Meeting - Agriculture - Brussels 22/23/24/25/26 February 1999 and 5/9/10/11 March 1999*. Press Release No. 6124/99. Council Secretariat, Brussels.

Council of the European Communities (1999b) Council Regulation (EC) No 1259/1999 of 17 May 1999 establishing common rules for direct support schemes under the Common Agricultural Policy. *Official Journal of the European Communities*, L 160.

Council of the European Union (2003) Council Regulation (EC) No 1782/2003 of 29 September 2003 establishing common rules for direct support schemes under the

Common Agricultural Policy and establishing certain support schemes for farmers. *Official Journal of the European Communities*, L 270, 1-69.

Daugbjerg, C. and Swinbank, A. (2004) The CAP and EU enlargement: prospects for an alternative strategy to avoid the lock-in of CAP support. *Journal of Common Market Studies* 42, 99-119.

European Commission (2001a) *The Agricultural Situation in the European Union. 1999 Report*. Office for Official Publications of the European Communities, Luxembourg.

European Commission (2001b) *Decision to Cut Red Tape for Small Farmers*. Press Release IP/01/882, 20 June 2001. European Commission, Brussels.

European Commission (2002) *Commission Publishes Indicative Figures on the Distribution of Direct Farm Aid*. MEMO/02/198, 1 October 2002. European Commission, Brussels.

European Council (1999) *Presidency Conclusions. Berlin European Council 24 and 25 March 1999*. Press Release. Council Secretariat, Brussels.

European Council (2002) *Presidency Conclusions. Brussels European Council 24 and 25 October 2002*. Press Release. Council Secretariat, Brussels.

Financial Times (2002) 28 October, 1.

Fischler, F. (2001) A three-pronged reform. Personal View. *Financial Times* 8 May, 19.

Guyomard, H., Bureau, J.C., Gohin, A. and Le Mouël, C. (2000) Impact of the 1996 US FAIR Act on the Common Agricultural Policy in the World Trade Organisation context: the decoupling issue. *Food Policy* 25, 17-34.

Harper Adams University and SAC (2002) *The Impact of Modulation*. A Study for The Royal Institution of Chartered Surveyors. RICS, London.

House of Commons (1997) *Hansard*. Written answers, Column 434, 31 July.

House of Commons (2001) *The Implementation of IACS in the European Union*. Agriculture Committee, Seventh Report Session 2000-01. HC 150. The Stationery Office, London.

Josling, T.E., Tangermann, S. and Warley, T.K. (1996) *Agriculture in the GATT*. Macmillan, London.

Josling, T.E. (2003) The current state of play in the WTO agricultural negotiations. *Agra Europe* 2055. 23 May 2003, A/1-A/2.

Kay, A. (1998) *The Reform of the Common Agricultural Policy. The Case of the MacSharry Reforms*. CAB International, Wallingford, UK.

Laird, S. (2002) A round by any other name: the WTO Agenda after Doha. *Development Policy Review* 20, 41-62.

MAFF (1999) *Arable Area Payments Scheme Explanatory Guide: Part I*. 2000 Edition. MAFF, London.

MAFF (2001) Memorandum submitted by the Ministry of Agriculture, Fisheries and Food. In: House of Commons Agriculture Committee, Seventh Report Session 2000-01. *The Implementation of IACS in the European Union*. Volume II Minutes of Evidence and Appendices. HC 150-II. The Stationery Office, London.

Meat and Livestock Commission (1993) *CAP Reform - One Year On*. MLC, Milton Keynes, UK.

Organisation for Economic Co-operation and Development (2002) *Agricultural Policies in OECD Countries Monitoring and Evaluation 2002 Highlights*. OECD, Paris.

Portuguese Government (2001) *A New Direction for European Agriculture. The Portuguese Contribution to CAP Reform.* Mimeo (see also *Financial Times*, 22 May 2001).

Roberts, I., Podbury, T., Freeman, F., Tielu, A., Vanzetti, D., Andrews, N., Mélanie, J. and Hinchy, M. (1999) *Reforming World Agricultural Trade Policies.* ABARE Research Report 99.12. Australian Bureau of Agricultural and Resource Economics, Canberra.

Ross, G. (1995) *Jacques Delors and European Integration.* Polity Press, Oxford, UK.

Sadoulet, E., de Janvry, A. and Davis, B. (2001) Cash Transfer Programs with Income Multipliers: PROCAMPO in Mexico. *World Development* 29, 1043-1056.

Schwaag Serger, S. (2001) *Negotiating CAP Reform in the European Union - Agenda 2000.* Report 2001, 4. Swedish Institute for Food and Agricultural Economics, Lund, Sweden.

Swinbank, A. (1993) CAP Reform, 1992. *Journal of Common Market Studies* 31, 359-372.

Swinbank, A. (1997) The New CAP. In: Ritson, C. and Harvey, D.R. (eds) *The Common Agricultural Policy*, 2nd Edition. CAB International, Wallingford, UK.

Swinbank, A. (1999) CAP reform and the WTO: compatibility and developments. *European Review of Agricultural Economics* 26, 389-407.

Swinbank, A. and Tanner, C. (1996) *Farm Policy and Trade Conflict. The Uruguay Round and CAP Reform.* The University of Michigan Press, Ann Arbor, Massachusetts.

Tangermann, S. (1998) An ex-post review of the 1992 MacSharry Reform. In: Ingersent, K.A., Rayner, A.J. and Hine, R.C. (eds) *The Reform of the Common Agricultural Policy.* Macmillan, London.

World Trade Organization (1999) Committee on Agriculture. *Notification from the European Communities.* G/AG/N/EEC/17. WTO, Geneva, Switzerland.

World Trade Organization (2000a) Committee on Agriculture, Special Session. *Note on Domestic Support Reform Negotiations on Agriculture. Submission from the United States.* G/AG/NG/W/17. WTO, Geneva, Switzerland.

World Trade Organization (2000b) Committee on Agriculture, Special Session. *EC Comprehensive Negotiating Proposal.* G/AG/NG/W/90. WTO, Geneva, Switzerland.

World Trade Organization (2000c) Committee on Agriculture. *Notification, domestic Support Commitments - Mexico - 1996, 1997 & 1998.* G/AG/N/MEX/7. WTO, Geneva, Switzerland.

World Trade Organization (2001a) Committee on Agriculture, Special Session. *Negotiation on WTO Agreement on Agriculture. Proposals by India in the Areas of: (i) Food Security, (ii) Market Access, (iii) Domestic Support, and (iv) Export Competition.* G/AG/NG/W/102. WTO, Geneva, Switzerland.

World Trade Organization (2001b) Ministerial Conference, Fourth Session, Doha, 9-14 November 2001. *Ministerial Declaration.* WT/MIN(01)/DEC/W/1. WTO, Geneva, Switzerland.

World Trade Organization (2002) *WTO Agriculture Negotiations. The Issues, and Where We Are Now.* Updated 25 January 2002, www.wto.org. WTO, Geneva, Switzerland.

Chapter Three:

Compensation Proposals for EU Farm Policy Reform

Jonathan Little[1], Thomas Knapp[2], Miguel Sottomayor[3] and Alan Swinbank[1]

Centre for Agricultural Strategy[1], The University of Reading, UK; Institut für Agrarökonomie[2], Georg-August Universität Göttingen, Germany; Faculty of Economics and Management[3], The Portuguese Catholic University, Portugal

Introduction

As noted in the Preface, there has been a long history of European agricultural economists advocating the deployment of decoupled payments to facilitate reform of farm policies. In this chapter, drawing on Beard and Swinbank (2001), we articulate some of the issues highlighted in the literature of the 1960s, 1970s and 1980s, prior to the advocacy of a bond scheme in 1990 in a report for the Land Use and Food Policy Inter-Group (LUFPIG) of the European Parliament by Tangermann (1990). Subsequently this report was published commercially (Tangermann, 1991). The chapter goes on to consider some of the issues that emerged in the ensuing debate on the LUFPIG study, but it is in Chapter 6 where a discussion of the fate of the Danish Government's 1992 proposal for a bond scheme is to be found. In Chapter 4 we re-articulate the component parts of a bond scheme and there, and in Chapter 7, we deal with the criticisms that have been raised on our initial working paper (Swinbank and Tangermann, 2000, 2001).

Defining decoupled support payments

The OECD (2000) has noted that decoupling 'is a general concept taken from the policy debate'. Indeed, it was the outcome of the Uruguay Round of GATT negotiations, with its notion of green, blue and amber boxes (as outlined in Chapter 2) that led to the current international interest in decoupling. Based upon the work of Cahill (1997), the OECD refers to *full decoupling*, which implies that decoupled payments do not influence either producer or consumer decisions: 'both the shape and position of the supply and demand curves should not be changed'. A less restrictive concept is that of *effective full decoupling* which results 'in a level of production and trade equal to what would have occurred if the policy were not in place'. A quota mechanism might, for example, be used to restrict the quantity supplied. The OECD has also attempted to develop a *degree of decoupling* index which would have a value of one when a policy is effectively fully decoupled, and of zero when the 'production and/or trade effects ... are equal to those of a PSE-equivalent increase in effective output prices' (OECD, 2000).

Roberts *et al.* (1999) describe decoupling as 'breaking the links between support and key market variables including production, prices and input use'. Decoupling should certainly encompass the notion that the link between compensation payments for the farm sector and current farming activities should be broken. However, the definition of decoupling as 'production-neutral income support' (Collins and Vertrees, 1988) is perhaps somewhat restrictive. Direct payments will raise the liquidity of farm households and that will have impacts on investment decisions. Changes to farm income, wealth and risk as a result of direct payments will inevitably influence production and create market distortions (Tielu and Roberts, 1998). Farmers may use the money to compensate losses in production to continue farming[1]. As Roberts and Andrews (1991) state, 'perfectly decoupled farm income support is virtually unattainable given that enhanced incomes of farmers are likely to affect their production decisions and capacity'. With this in mind, perhaps a more sensible definition of 'decoupling' might refer to policies which do not *directly* affect production:

> 'The idea is a simple one. The payments should not be related to current output decisions. The payments to resources would be based on projections of losses due to the reduction of protection' (Johnson, 1991).

By decoupling in this way, 'policies can be designed which have much smaller impacts on production, consumption, trade and world prices than do many of the policies which are currently in use. Although support would still be provided, it would be largely trade neutral' (Roberts and Andrews, 1991).

Early proposals for decoupled payments

Nash (1961), a British agricultural economist, was a firm believer that the farm sector 'must operate under the ordinary rules of solvency and under a price system which ceases to misdirect the efforts and energies of producers by distorting the incentives governing their actions'. However, he conceded that 'the initial economic loss which agriculture would face from a general reversal of the existing protective arrangements would be severe'. Thus, he argued, there would be a case for shielding 'the industry from the immediate losses that would result' provided such 'payments do not influence the productive decisions of those who remain in agriculture or induce anyone to remain in preference to retiring or finding another occupation'. He explained that:

> 'the simplest method of compensating agriculture for the withdrawal of the protective system would be an unconditional payment to all those at present engaged in farming, or to those of them deemed to be in need of compensation, calculated by reference to the difference between the incomes now earned under the protective system and those capable of being earned under a system of free market prices. An annuity calculated in this way and payable for life to all engaged in farming, but not transferable to their successors, would, in theory at least, make it possible to bring the protective system to an end while fully making good the loss of income to its present beneficiaries' (Nash, 1961).

In an earlier article, in *The Guardian*, he had, however, suggested that 'it might be desirable deliberately to under-assess the amount of the payment and to provide at the same time for additional benefits to be available on appeal to producers who could prove that the change had reduced their net income' (Nash, 1960). Furthermore:

> 'The payment of these additional benefits could if necessary be made subject to a means test, and they could take such forms as the right in approved cases to receive the capital value of the annuity in a lump sum to be used to equip the applicant to make a start in some new business outside agriculture, the right to some increase in the annuity conditional upon the retirement of elderly applicants from farming, or the right in suitable cases to use the capital value of the annuity in stocking a larger farm' (Nash, 1960).

Thus, farmers would either receive an annual compensation payment, *or* a capital sum reflecting the present value of the flow of annual compensation payments to which the farmer was entitled. This would have involved the government fixing the interest rate that would be used to determine the present value of the future income stream, as well as fixing the appropriate level of compensation. Furthermore, if the government is expected to fund an up-front capital sum, rather than an annual flow of payments, the annual profile of budget expenditure is clearly different.

In 1970, two panels of experts advocated reductions in CAP support prices, offset by compensatory payments. An Agricultural Study Group of the Federal Trust for Education proposed a series of 'auxiliary payments' that:

> 'should be temporary, ... should not represent an indirect stimulus to production and ... should advance the adjustment of agriculture so that by the time they are withdrawn farmers will be able to secure an adequate income from the market without either high consumer prices or price subsidies' (Marsh, 1970).

An expert panel established by the Paris based Atlantic Institute, in proposing a reduction in CAP support prices, commented that 'It is out of the question to propose such price adjustments without compensatory measures in the form of direct aids, which will mitigate their effect on income, and without effort to establish more normal and stable prices in international markets' (Uri, 1970). The compensation payments would have had four characteristics:

> 'The first of these would compensate for lower prices on the basis of average yield in the Community. The second would fix payments depending upon the area under cultivation *in the years preceding* the decision to lower prices. The third would involve minimum and maximum time periods and would limit the subsidy to the life of the farm operator, so that it could not be passed on to his successors. The fourth would make it vary inversely with the size of farms' (Uri, 1970).

Josling (1974) cited the Atlantic Institute report as an example of policy advice in the 1960s in which 'economists advised direct income payments unrelated to output as a way of ensuring reasonable standards of living to rural people if such were not forthcoming through the sale of their produce'.

Koester and Tangermann (1977) later suggested that the reason direct income payments independent of production volume had not been accepted might, in part, be due to the fact that they involved dramatic shifts in policy. 'The political tendency to adopt stepwise procedures' led them to advocate the retention of price policy but to also recommend that 'farm prices should decrease - in real terms - by 2 to 2.5% p.a. The loss in sector income ... should be compensated by personally tied income payments. These payments should be made to the farmer himself, independent of how many workers are employed on the farm at present and in future and also independent of the future production pattern' (Koester and Tangermann, 1977). Their plan prevented future generations from inheriting the direct payments but did allow for their capitalisation, if farmers migrated out of agriculture and some other conditions were fulfilled.

The Koester and Tangermann (1977) proposal to 'grant the income payments - not at EC level but nationally (i.e. per Member State) and also to finance them nationally' was seen by Meester (1980) as a very important disadvantage. He felt that the proposal 'entails the danger of an increasing distortion of competition and an accompanying ultimate disintegration of the common agricultural market'. He concluded that a system of national income payments would be more attractive to wealthier Member States, such as West Germany, but that, ultimately, it could result in a return to 'the national protection of agriculture as it existed before the beginning of the CAP'. This is also a favoured topic of Portuguese and Spanish authors, who generally claim that any move to re-nationalise agricultural aid, which they oppose, would jeopardise the EU's proclaimed objective of cohesion and would introduce market distortions inside the EU, favouring the better structured and better endowed agricultural regions (Azcarate, 1996; Lourenco, 1996; Vinas, 1996).

At the same time as Meester (1980) was expressing his concerns over plans for the decentralisation of payments from the CAP, Priebe (1980) was adding his support for decoupled income subsidies in a paper that looked for 'a solution to the conflict between incomes and market policies'. Priebe (1980) stated that 'In order to reduce the price stimulus to increased production, income support must be granted independently of production'. He went on to suggest that the preferred criterion for allocating income subsidies should be the surface area farmed rather than herd size, as this could artificially obstruct a move towards more extensive farming.

In 1980, Bergmann, a member of the Atlantic Institute's expert panel of 1970, again sought to highlight the desirability of direct income payments by underlining the considerable academic interest in the topic throughout the decade since the Atlantic Institute report. He described these types of payments as being 'at least among economists, the most popular method for alleviating the income effect of lower support prices' (Bergmann, 1980). The eligibility criterion for the payments would be steered away from current production decisions and based instead on an historic time frame. Again, Bergmann's (1980) proposal suggested that eligibility for payments to farmers 'should not be transferred to their children except if they were to die shortly after the program started'. His proposals also included a ceiling on the amount any one farm could receive as well as the suggestion that the administrative cost of dealing with the very smallest of farms might make their exclusion a regrettable necessity.

Marsh (1981) was keen to point out that, when compensating farmers for any proposed price reductions 'it was important to differentiate between the loss of revenue and loss of income'. He felt that in the longer run 'such compensation should be phased out for those farmers who are able to adapt. For the remainder, the old, the small-scale and the farmer in very harsh conditions, compensation might be virtually permanent. In such cases it would be, in effect, a pension'. He also underlined the importance of distinguishing the 'welfare' reasons for compensation, which must be limited to income, and the 'adjustment' reasons. 'One way to cope with the adjustment problem would be to allow farmers to take the capitalised value of compensation payments as a lump sum at the outset' (Marsh, 1981).

Harris *et al.* (1983) concluded that it was probably unrealistic to believe that substantial price cuts could be achieved without some degree of income compensation. They commented that the payment to an individual should not be dependent upon the individual's future contribution to farm output, but noted that an endless variety of schemes could still be envisaged. They asked: 'how is the target group to be defined, how is the income loss to be measured, how long are the payments to continue, how are the payments to be adjusted for inflation, and is there to be a maximum annual payment per person (or family)?'

Clearly, in terms of the policy space depicted in Fig. 1.1 on page 2, many of the proposed schemes lay somewhere on the line connecting the apex of the triangle (compensation) with the bottom right-hand corner (income support): the perceived need to maintain reasonable standards of income for rural people was clear. This is reflected in the frequently encountered suggestion that payments should be limited to the lifetime of the original recipient: compensation is proposed not to protect the farm business, but instead the farmer's income. Not all the literature published at this time, however, was supportive of direct or decoupled payments in pursuit of this goal. Weinschenck (1975) was particularly troubled by the difficulties involved in defining the eligibility criteria for payments of this kind:

> 'To draw the separation line between receivers and non-receivers of direct income payments is an almost impossible task for the Common Market institutions, since a common criterion which is acceptable for all countries and for the Community as a whole can hardly be found' (Weinschenck, 1975).

Munk (1989) questioned the assertion that the lump sum transfers were more efficient than traditional mechanisms for supporting farm incomes. He felt it was particularly important to 'recognize important aspects of reality which are not captured by the standard welfare economic models'. In particular he highlighted the need for lump sum transfers to be financed out of 'distortive tax instruments', a problem recognised in the public finance literature.

Furthermore, he suggested the accompanying restructuring of the industry would involve potential for rough justice:

> 'On the expenditure side, in order for a system of income transfers to the farmers to be non-distorting, the payments must be independent of the farmers' choice of occupation. This means that the budget costs of a system of lump sum transfers in the EC could be considerable, at least in the short term, if it were to provide the same income to farmers as the present system of price support. Not only the farmers who would stay in the agricultural sector, but also the farmers who would leave, would have to receive the lump sum transfer in order for such a system to be non-distorting. Under the present system of price support a lot of farmers have left the EC agricultural sector without receiving compensation. ... Under a realistic system of lump sum payments a lot of people would receive payments which would be far in excess of what was needed to compensate them for leaving the agricultural sector. On the other hand, there would be people who would be under-compensated. ... One may therefore safely assume that solving the problem of low farm income in the EC cannot be done by lump sum transfers' (Munk, 1989).

Munk and Thomson (1994) later noted that if lump sum transfers were to be 'targeted', then they 'are associated with high 'transaction costs' because in order to achieve their objectives, i.e. to increase the income of those with low income or those who deserve compensation, such transfers need to be based on detailed information about individual households'. Munk (1989) did, however, concede that the evolution of modern information technology that made the processing of information much easier meant that 'lump sum payments based on farmers' past production seem relatively less costly and partly, therefore, politically more feasible today than in the 1960s'.

Roberts and Andrews (1991) raised another concern when they claimed that resource misallocation would remain if decoupled policies were introduced for a restricted group of agricultural products, whilst distorting policies were retained for others with which the newly decoupled products are competing for inputs. In some instances the overall misallocation could even be increased as additional resources are channelled into producing supported products to which coupled policies still apply:

> 'It is very difficult to provide effective, decoupled lump sum support on a commodity-specific basis. The objective of lump sum support is to raise incomes of farmers rather than incomes of producers of particular items and it is likely to be more effectively decoupled if it is provided generally to all who meet specified definitional criteria for being farmers' (Roberts and Andrews, 1991).

The bond scheme

In a report for the LUFPIG of the European Parliament, Tangermann (1990) recommended that farmers be issued with bonds on which the Community would make annual payments for a certain number of years, in order to compensate for cuts in support prices[2]. Tangermann's paper had evolved from a cost-benefit analysis of alternative farm policies for German agriculture produced for the German Ministry of Agriculture (Koester and Tangermann, 1977).

> 'The total amount of bonds would be based on the income loss expected to result from the cut in support prices. Bonds would be allocated to individual farmers on the basis of their output in a recent reference period, and future production decisions would not affect the amount of bonds. Bonds would be transferable and could be sold on the private capital market' (Tangermann, 1990).

With such a scheme, payments would be made to the bond-holder, and would not be conditional upon future farming activities. Therefore, farmers could decide whether they wanted to retain their bonds in order to receive annual payments, or whether they wanted to sell them on the capital market in order to have a capital sum that may help them relocate or start a new occupation. A farmer could even retire and still receive the annual payments. In this way, the bond enables farmers to be compensated for the removal of price support and to adjust to more competitive market conditions, but does not seek to encourage farmers to remain in production[3].

Tangermann suggested a 15 year life-span for the bond. It was important that the bond payments would only be made for a limited duration, with the major argument for granting compensation being that it takes time for farmers to adjust to changes in policy. Once all the adjustments have been made, no further compensation is warranted:

> 'If the compensation were made permanent, the intended effects of the policy change would be vitiated, as new generations of farmers would be 'artificially' attracted into farming through the promise of permanent income supports not available to other sectors of the economy' (Marsh, 1991a).

One advantage of a bond scheme is that annual payments can (in part) be financed out of savings resulting from price cuts. The farmer's ability to sell the bond for a capital sum does not imply an up-front call on the EU's budget. It is the financial markets which would provide the capital sum, whilst the EU's Budget would be called upon to make regular annual payments to the bond-holders - whoever they might be - for the lifetime of the bond.

Marsh (1991b) conceded that political pressure could result in a compensation package related to farm size. It might be expected that compensation should be more generous to relatively small than to relatively large farmers since the small farmer has few alternative opportunities for his land and his labour whereas the large farmer is better equipped to adapt to change: 'One can dispute the economics of such discrimination but the politics of it seem to me essential'.

Implementation of a bond scheme would require decisions on, for instance, the overall amount of compensation payments and their duration, the degree to which payments are modulated with increasing farm size, and the exact criteria used for the distribution of bonds between farms. The LUFPIG Report recognised that the costs of a bond scheme would depend on such political decisions. It gave an example of the scale of costs by reference to the proposed GATT agreement to reduce support by 30%. The report assumes that this would have reduced farm incomes by about 15% in real terms between 1990 and 1995. If all commercial farmers were to be compensated for such an income loss, modulated by farm size with the compensation for larger farmers fixed at 60% of the full rate, the annual cost of payments would be roughly 7.1 billion ecu. If all farms were to be compensated in an equivalent manner, total annual payments would have to be around 10 billion ecu (Tangermann, 1991).

Tangermann's bond scheme was enthusiastically endorsed by Sir Leon Brittan, then an EU Commissioner. In highlighting the need for a solution that was 'at once radical and realistic', he praised the idea 'that income aid should go to farmers in the form of a saleable 'bond'' (Brittan, 1994):

> 'Such a bond would have a healthy impact economically and psychologically: each farmer would know the aid was finite, focussing his mind on the need to gear himself for a new career; and it would give him the financial backing to find new work, and in good time, he could choose if and when to leave the land, selling the bond in order to tide him over until he was established in a new line of business. It would be a one-off lump-sum, which although costly would put an end to increasing support measures in the future and bring no new incentives to increase

farm output. Instead, it would help heal the sclerosis currently gripping Europe's agricultural markets: it would loosen up trade in farm property, lowering land prices as more farmers began buying and selling 'Tangermann bonds', making it easier for new farmers to start up in viable areas of agricultural production. The sheer finality of it, although hard for farmers to digest at first, would in fact encourage them and their governments to tackle Europe's farming future today, before it is too late' (Brittan, 1994).

The proposals of Tangermann and the LUFPIG panel were also backed wholeheartedly by a report from the UK House of Lords Select Committee on the European Communities. The Committee concluded that 'a fixed redeemable bond ... appears a more promising means of providing compensatory or adjustment aid to farmers', with the then UK Minister of Agriculture, John Gummer, saying that he thought it 'rather an ingenious scheme. ... I think that it has a lot of intellectual attraction' (House of Lords, 1991). A year later, the Consumers in the European Community Group cited Tangermann's plan when advocating 'a shift away from price supports to direct aids only where needed' (Consumers in the European Community Group, 1992).

In refining his proposals in the light of the MacSharry reform package, Tangermann attempted to tackle several contentious issues. When considering whether compensation payments should be made to farmers or to farms, he concluded that making them to farms rather than to farmers would make it impossible to turn compensation into bonds and that it 'effectively means to link payments to land'. 'Future payments would then be capitalised in the value of land. The resulting higher land prices would again retard structural adjustment in agriculture' (Tangermann, 1992).

Another issue which proved harder to resolve concerned the allocation of payments between landlords and tenants. Tangermann concluded that 'this is a highly complex and also extremely sensitive issue, and some aspects touch upon equity considerations which are beyond economic analysis' (Tangermann, 1992). Swinbank believed that the introduction of the bond would result in a heavier fall in the value of landlords' capital than in tenants' capital. But, in referring to the precedent of milk quota, or entitlement to ewe premia, being sold off the farm, he concludes:

'For both milk quota and ewe premia, the British legal system has found mechanisms for the appropriate allocation of quota value between landlord and tenant, and so farm ministers should perhaps be encouraged to press ahead with this long overdue reform without worrying unduly about the eroding asset values of landowners' (Swinbank, 1997).

In 1993, Poole drew on Tangermann's original plan to develop his own proposal for reforming the CAP with the use of *exit bonds*. One novel element to his scheme was that a choice of bond schemes would be offered to farmers. They could accept either a bond on which an annual payment was made, with zero redemption value ('an 'annuity' type income only instrument paying interest in the form of a coupon for a fixed term'), or a 'zero coupon (deep discount) instrument' which would not yield a regular income stream, but would pay-out a fixed capital sum on maturity (Poole, 1993). Perhaps more significantly, Poole's bond scheme would be introduced on a voluntary basis. He envisaged his proposed scheme operating in this way: 'farmers would be given the option of receiving an amount equal to this payment stream (the area and headage payments under the MacSharry Reforms of 1992) in the form of a financial instrument guaranteed by the EC which would be freely tradable on a secondary market. By taking such an instrument, the farmer and the land farmed would become ineligible for additional agricultural price or income support'. Presumably there would be a one-off opportunity to accept a bond, and then the farmer's decision to accept or decline the bond would be irreversible.

Harvey (1997) also alluded to the possibility that the bond could be voluntary. Such a reform, Harvey continued, would 'save the bureaucracies considerable and ongoing implementation and policing costs, which would be incurred under the bond option on a once-and-for-all basis'. However, these savings on bureaucracy are probably unlikely if the bond was indeed to be made voluntary, with the existing administrative system being required for those farmers who chose not to take the bond. James Provan, MEP, in another LUFPIG publication, had stated that one of the bond scheme's most attractive features was 'that it would help reduce bureaucracy', the reduction of which, he felt, should be 'the litmus test of any further reform of the CAP' (Provan, 1996). Harvey (2000) and others (for example Sturgess, 1998 and Thurston, 2002) have continued to canvass the advantages of a bond scheme. In Chapter 5 we discuss the proposal for a Common Agricultural and Rural Policy for Europe (CARPE) (Buckwell *et al.*, 1997) which would include decoupled Transitional Adjustment Assistance (TAA) payments.

Commission initiatives

In the past, the EU has deployed a number of limited schemes to encourage the early retirement of farmers, to speed the process of farm amalgamation, or to grub-up orchards and vineyards and cease or suspend production of milk (see Fennell, 1997). The reform of agriculture envisaged in the *Mansholt Plan* was based on measures to remove surplus labour of all ages from farming, the creation of farm enterprises of an adequate economic size, the improvement of the operation of markets, and the adjustment of supply to demand. In order to achieve the desired results, the plan envisaged that it might be necessary to provide a personal aid to farmers who would not be in a position to benefit from the modernisation and retirement aid being offered. These personal aids would have been independent of the volume of production and the inputs on the farm and would have been granted within limits set by the regional situation and the age of the farmer, or farm worker, involved (Fennell, 1997). In terms of the triangle of Fig. 1.1, they lie towards the right-hand corner.

One of the European Commission's most detailed policy statements on direct income aids was made in its *Perspectives for the Common Agricultural Policy* (Commission, 1985). In this policy reform document, the Commission laid out, for discussion, four different options for income aids:

Option A: Pre-pension for farmers of 55 years and older;

Option B: A structural approach;

Option C: A social approach;

Option D: A buying out approach.

The second option, of a structural approach, was particularly interesting in terms of the specification of the aid payments:

> 'The aid would be *temporary* (e.g. limited to a 5 year 'period of transition'), giving the farmer a financial relief during some years in order to allow him to decide on his future and to make the necessary adjustments. Furthermore in order to avoid too abrupt a cut-off at the end of the transitional period the aid would need to be *degressive*' (Commission, 1985; emphasis as in the original)[4].

The Commission also stressed that 'great care would have to be taken to keep, as far as possible, such schemes neutral with respect to production and compatible with market policy' (Commission, 1985). This echoed a view of the Commission ten years earlier when it stated that income aid 'must not be linked with specific types of production: so long as the beneficiary continues to practise farming it is preferable that he should concentrate on those products in relation to which - at market prices - his productivity is highest' (Commission, 1975).

As noted in Chapter 1, in 1989 the EU did introduce a limited scheme of income aids (Council, 1989). But, as Fennell (1997) notes, in 1992 the Commission reported that the number of farmer recipients was only 180,000, far short of the 850,000 that had been expected. Not all Member States took up the option of applying the scheme, and only The Netherlands explicitly linked the amount of aid to the revenue loss brought about by the 1988 CAP reforms.

On two occasions the Commission has proposed a bond scheme. In 1991, as we saw in Chapter 2, it proposed a bond scheme for the dairy sector; but this was not endorsed by the Council. More recently, in a report on the common organisation of the market in raw tobacco, the Commission discussed several rejected scenarios, before identifying its preferred option. Alongside modulated premia, the Commission proposed a bond scheme to assist farmers to leave the sector voluntarily:

> 'This aid could take the form of buying up quotas over a sufficiently long time-span (7-10 years) coupled with a degressive payment mechanism or, alternatively, issuing the affected tobacco farmers bonds, on the basis of which the Community would pay out annual annuities until the bond's maturity (7-10 years). The tobacco

farmers would be free either to keep the bonds and collect the corresponding annuities or sell them on the private market' (Commission, 1996).

As it turned out, the quota buy-back option was introduced and another opportunity for the adoption of a bond scheme had been lost.

Notes

[1] Although only anecdotal, the initial comment of Daniel McFadden of the University of California, Berkeley, on learning of his award (jointly with James Heckman) of the Bank of Sweden Prize in Economic Studies in Memory of Alfred Nobel, that 'I'm just going to keep farming until the prize is gone' has resonance (as quoted in the Financial Times, 12 October 2000). McFadden, whose work focuses on how individuals and households make choices, has a hobby farm producing wine grapes in the Nappa Valley in northern California. It would appear that he soon had second thoughts, describing the farm as a 'money losing business' (College News from University of California, Berkeley, dated 11 October 2000: http://Is.berkeley.edu/new/00/macfadden.html), and donated the prize to the East Bay Community Foundation for the promotion of arts and education (http://www.nobel.se/economics/laureates/2000/mcfadden-autobio.html accessed 26 September 2002).

[2] Thinking again of the triangle depicted in Fig. 1.1, it is, perhaps, unfortunate that the chapter was headed 'A bond scheme for supporting farm incomes' (Tangermann, 1991). Tangermann quite clearly advocated decoupled compensation payments, paid over a limited time period.

[3] Bonds have been issued to compensate economic actors in a number of situations. Tony Byrne has pointed out to us similarities between the proposed bond scheme and a system of compensation introduced in the late nineteenth century in Japan for the Samurai, after the Meiji restoration of 1868, buying the Samurai out from their stipends. In a number of Latin American states, long-term bonds (often non-transferable) were given to former landowners whose land had been expropriated in the land reforms (Barraclough and Collarte, 1973). When swathes of British industry were nationalised after the Second World War, the previous owners often received compensation bonds which had gilt-edged status because interest payments were guaranteed by the Treasury (Robson, 1962).

[4] Degressive, in this context, means declining over time. This is its usual meaning in this literature, but the reader needs to be cautious. For example, Tangermann (1991) used the term 'degressivity' to refer to the possibility that the degree of compensation would fall as farm size increased, with only the smaller farms receiving full compensation; and similarly The Atlantic Institute report referred to a degressive element in its proposal (Uri, 1970). More recently the word

'modulation' has been used in 'CAP-speak' to refer to this latter concept, although 'modulation' can itself have different meanings according to context.

References

Azcarate, T.G. (1996) La politica agraria comun a debate. *Revista Espanola de Economia Agraria* 176-177(2-3), 243-264.

Barraclough, S. in collaboration with Collarte, J.C. (eds) (1973) *Agrarian Structure in Latin America.* Lexington Books, Lexington, Massachusetts.

Beard, N. and Swinbank, A. (2001) Decoupled Payments to Facilitate CAP Reform. *Food Policy* 26, 121-145.

Bergmann, D. (1980) Possible alternatives to the CAP and their economic consequences. Paper at a conference of the Agricultural Economics Society on *The CAP in the 1980s* at The Royal Society, London.

Brittan, L. (1994) *Europe. The Europe We Need.* Hamish Hamilton, London.

Buckwell, A., Blom, J., Commins, P., Hervieu, P., Hofreither, M., von Meyer, H., Rabinowicz, E., Sotte, F. and Sumpsi Vîna, J.M. (1997) Towards a common agricultural and rural policy for Europe. *European Economy* Reports and Studies No 5. CEC, Luxembourg.

Cahill, S.A. (1997) Calculating the Rate of Decoupling for Crops Under CAP/Oilseeds Reform. *Journal of Agricultural Economics* 48, 349-378.

Collins, K. and Vertrees, J. (1988) Decoupling and U.S. Farm Policy Reform. *Canadian Journal of Agricultural Economics* 36, 733-745.

Commission of the European Communities (1975) *Stocktaking of the Common Agricultural Policy.* COM(75)100. CEC, Brussels.

Commission of the European Communities (1985) *Perspectives for the Common Agricultural Policy.* COM(85)333. CEC, Brussels.

Commission of the European Communities (1996) *Report from the Commission to the Council on the Common Organization of the Market in Raw Tobacco.* COM(96)554. CEC, Brussels.

Consumers in the European Community Group (1992) *The Future of the Common Agricultural Policy.* Briefing paper. CECG, London.

Council of the European Communities (1989) Council Regulation (EEC) No 768/89 of 21 March 1989 establishing a system of transitional aids to agricultural income. *Official Journal of the European Communities,* L 84.

Fennell, R. (1997) *The Common Agricultural Policy. Continuity and Change.* Clarendon Press, Oxford, UK.

Financial Times (2000) 12 October, 10.

Harris, S., Swinbank, A. and Wilkinson, G. (1983) *The Food and Farm Policies of the European Community.* John Wiley & Sons, Chichester, UK.

Harvey, D.R. (1997) The CAP in the 21st Century. In: Ritson, C. and Harvey, D. (eds) *The Common Agricultural Policy,* 2nd edition. CAB International, Wallingford, UK.

Harvey, D.R. (2000) The future of the CAP? Problems and Proposals. *Economic Affairs* 20(2), 34-41.

House of Lords (1991) *Development and Future of the Common Agricultural Policy. Volume I.* Sixteenth Report from the Select Committee on the European Communities, Session 1990-1991. HL 79-I. HMSO, London.

Johnson, D.G. (1991) *World Agriculture in Disarray,* 2nd Edition. Macmillan, London.

Josling, T.E. (1974) Agricultural policies in developed countries: a review. *Journal of Agricultural Economics* 25, 229-264.

Koester, U. and Tangermann, S. (1977) Supplementing farm price policy by direct income payments: cost-benefit analysis of alternative farm policies with a special application of German agriculture. *European Review of Agricultural Economics* 4, 7-31.

Lourenco, J. (1996) *Reforma da PAC, Politica de Qualidade e Alargamento da Uniao Europeia,* Working Paper No 4. Instituto Superior de Agronomia, Departamento de Economia Agraria e Sociologia Rural, Lisbon.

Marsh, J.S. (1970) *A New Agricultural Policy for Europe.* Proposals submitted by the Agricultural Study Group of the Federal Trust. Federal Trust for Education & Research, London.

Marsh, J.S. (1981) The response of the CAP to the needs of Europe in the 1980s. In: Marsh, J.S., Mahé, L.P. and Roudet, M. (eds) *The CAP in the 1980s: Two Views on Revision.* Centre for European Agricultural Studies, Wye College, Ashford, Kent, UK.

Marsh, J.S. (1991a) The requirements for the CAP in the 1990s. In: Marsh, J.S., Green, B., Kearney, B., Mahé, L., Tangermann, S. and Tarditi, S. (eds) *The Changing Role of the Common Agricultural Policy: the Future of Farming in Europe.* Belhaven, London.

Marsh, J.S. (1991b) Minutes of evidence. In: House of Lords *Development and Future of the Common Agricultural Policy. Volume II. Minutes of Evidence.* Sixteenth Report from the Select Committee on the European Communities, Session 1990-1991. HL 79-II. HMSO, London.

Meester, G. (1980) The future of the CAP: a Dutch view of the policy options. In: Smith, P.F. (ed) *Alternative Proposals for the Common Agricultural Policy. A Symposium of Views.* Irish Section, European League for Economic Cooperation, Dublin.

Munk, K.J. (1989) Price support in the EC agricultural sector: an optimal policy? *Oxford Review of Economic Policy* 5(2), 76-89.

Munk, K.J. and Thomson, K. (1994) The economic costs of agricultural policy. *European Economy.* Reports and Studies No 4. CEC, Luxembourg.

Nash, E.F. (1960) Farming with free prices: towards a competitive agriculture. *The Guardian* 18 August.

Nash, E.F. (1961) Agriculture. In: Seldon, A. (ed) *Agenda for a Free Society.* Hutchinson, for The Institute of Economic Affairs, London.

Organisation for Economic Co-operation and Development (OECD) (2000) *Decoupling: a Conceptual Overview.* Directorate General for Food, Agriculture and Fisheries, COM/AGR/APM/TD/WP(2000)14/FINAL. OECD, Paris.

Poole, P. (1993) *Reform of the Common Agricultural Policy: the Use of Exit Bonds.* European Policy Forum, London.

Priebe, H. (1980) Towards a solution of the conflict between incomes and market prices. In: Smith, L.P.F. (ed) *Alternative Proposals for the Common Agricultural Policy. A*

Symposium of Views. Irish Section, European League for Economic Cooperation, Dublin.

Provan, J. (1996) *Europe's Freedom to Farm. The Future of the Common Agricultural Policy: Risk Management in the Marketplace*. Land Use and Food Policy Inter-Group, European Parliament, Brussels.

Robson, W.A. (1962) *Nationalized Industry and Public Ownership*, 2nd Edition. George Allen and Unwin, London.

Roberts, I.M. and Andrews, N.P. (1991) Decoupling and the 1990 US farm bill for grain. *Agriculture and Resources Quarterly* 3, 203-219.

Roberts, I., Podbury, T., Freeman, F., Tielu, A., Vanzetti, D., Andrews, A., Mélanie, J. and Hinchy, M. (1999) *Reforming World Agricultural Trade Policies*. ABARE Research Report 99.12. Australian Bureau of Agricultural and Resource Economics, Canberra.

Sturgess, I. (1998) The future of the Common Agricultural Policy. *Royal Institute of Chartered Surveyors Research Conference: ROOTS '98*. RICS, London.

Swinbank, A. (1997) The New CAP. In: Ritson, C. and Harvey, D.R. (eds) *The Common Agricultural Policy*, 2nd Edition. CAB International, Wallingford, UK.

Swinbank, A. and Tangermann, S. (2000) *A Proposal for the Future of Direct Payments under the CAP*. Working Paper No 1 of the Bond Scheme Project. Centre for Agricultural Strategy, The University of Reading, Reading, UK.

Swinbank, A. and Tangermann, S. (2001) The Future of Direct Payments under the CAP: a Proposal. *EuroChoices* Premier Issue, Spring, 28-29, 32-34.

Tangermann, S. (1990) A bond scheme for supporting farm incomes. In: Marsh, J.S., Green, B., Kearney, B., Mahé, L., Tangermann, S. and Tarditi, S. (eds) *A Future for Europe's Farmers and the Countryside*. Land Use and Food Policy Inter-Group. European Parliament, Brussels.

Tangermann, S. (1991) A bond scheme for supporting farm incomes. In: Marsh, J.S., Green, B., Kearney, B., Mahé, L., Tangermann, S. and Tarditi, S. (eds) *The Changing Role of the Common Agricultural Policy: the Future of Farming in Europe*. Belhaven, London.

Tangermann, S. (1992) *Reforming the CAP? In For a Penny, In For a Pound*. IEA Inquiry No 28. Institute of Economic Affairs, London.

Thurston, J. (2002) *How to Reform the Common Agricultural Policy*. European Rural Communities Paper 1. The Foreign Policy Centre, London.

Tielu, A. and Roberts, I. (1998) Farm income support: implications for gains in trade of changes in methods of support overseas. *ABARE Current Issues* 98(4), 1-8.

Uri, P. (1970) *A Future for European Agriculture*. A Report by a Panel of Experts. The Atlantic Institute, Paris.

Vinas, J.M.S. (1996) La agricultura espanola ante los nuevos escenarios de la PAC. *Revista Espanola de Economia Agraria* 176-177(2-3), 265-301.

Weinschenck, G. (1975) Basic alternatives of the future agricultural policy in the European Common Market. *Journal of Agricultural Economics* 25, 145-158.

Chapter Four:

A Bond Scheme to Facilitate CAP Reform

Alan Swinbank[1] and Stefan Tangermann[2]
Centre for Agricultural Strategy[1], The University of Reading, UK; OECD[2], Paris, France

Introduction

The purpose of this study was to explore the practicalities and acceptability of a bond scheme as *part of* CAP reform. To this end, it was necessary to re-articulate, and clarify, the characteristics of a bond scheme. We did this in Swinbank and Tangermann (2000, 2001), and this chapter is derived from that earlier work. This chapter, then, makes suggestions for a further step in CAP reform: it *advocates* a particular policy response. As detailed elsewhere in the text, in July 2002 the Commission proposed, and in June 2003 the Council accepted, a partial decoupling of direct payments from production. Thus the bond scheme could be seen *either* as a more radical alternative to the European Commission's proposals for the mid-term review, *or*, following on from the Fischler reforms of 2003, as the next stage in CAP reform.

Our primary focus had been the future of the direct payments made under the CAP, as introduced as part of the 1992 (MacSharry) reform and extended under Agenda 2000. We emphasised that this policy change would need to be set in the context of a wider redirection of the CAP to reflect the environmental and rural development concerns of society (the 'non-trade concerns' referred to in the WTO *Agreement on Agriculture*, related to the multifunctionality of agriculture, and the second pillar of the CAP articulated in the Agenda 2000 debate) and, in Chapter 5, we expand on this theme. Nonetheless, the main objective of the proposal was to create certainty about the future of the existing system of direct payments, in a situation in which farmers were increasingly concerned about the fate of these payments. The policy change would also contribute to raising farm incomes and improving the efficiency of the EU economy.

The policy change advocated is to convert current and future direct payments into bonds that will give their holders a guaranteed future stream of payments.

These bonds could be sold on the capital market, generating a capital sum for the original recipients of the bonds. Such a scheme would completely decouple payments from production and resource use in agriculture, and thereby improve greatly the functioning of market forces and the competitiveness of EU agriculture.

We believe that the payments introduced by the MacSharry reform of 1992 (and augmented in Agenda 2000) to compensate farmers for the cuts in support prices made in the process of CAP reform, and to help farmers adjust to the consequences, and now carried forward into the Single Payment Scheme by the Fischler reforms, were justified at the time. Farmers had to be given an opportunity to adjust to changes in policies that they could not foresee.

These area and headage payments had become a major part of support provided under the CAP and, on the basis of these payments, farmers had made investments, and indeed taken decisions on their future professional activities. Some of these decisions must be revised as the policy is changed. As such adjustments take time, policy makers must consider the need for adjustment assistance when policy deviates from the track established in the past.

Since compensation payments introduced as a consequence of policy reforms have only this objective, we believe that such payments should not be conditional on any other requirements, such as the environmental effects of farming practices. There may well be reasons for also making payments to reward services that farmers provide to society in general, such as maintaining the countryside or improving the rural environment, and we discuss this issue more fully in Chapter 5. However, in this chapter our primary concern is not with payments for services rendered, but rather with compensation payments made as part of reforms of past market and price policies. Such payments allow farmers time to adjust to a new policy and market environment. They might indeed be termed adjustment payments.

The chapter starts by discussing the need for taking decisions on the future of direct payments. It then describes the implementation of a bond scheme, by outlining six component parts that would in their entirety transform the current payment regime into a bond scheme. The Single Payment Scheme has begun this process. By adopting this stepwise description, it is hoped that the objectives and the nature of a bond scheme will be better understood. After a brief look at some major advantages of a bond scheme, the chapter then introduces some of the major issues that have figured prominently in the debate about the introduction of a bond scheme; more detailed implementation issues will be discussed in greater detail in Chapter 7.

Why is there urgent need to decide on the future of direct payments?

Since the MacSharry reform of 1992, direct payments have become a central element of the CAP. As detailed in Chapter 2 (see Table 2.3), more than half of

CAP expenditure (EAGGF Guarantee) now goes into direct payments. On average, across all farms in the EU, these direct payments are equivalent to about 20% of net value added[1]. For a policy instrument as important as this, it is imperative that its future fate is made crystal clear. Farmers need to know the basis on which they should make their investment plans. Policy makers have to build plans for future payments into the EU's budget forecasts, and into the design of all other CAP measures.

However, when direct payments were made a central element of the CAP in 1992, no explicit decisions on their future were taken. Although no time limit was set, many farmers have always harboured doubts about the long-run sustainability of the payments. These doubts were reinforced during the Agenda 2000 debate when the governments of some Member States proposed that payments should be made degressive over time. The decisions taken at Berlin in March 1999 did not contain that element, and in their own way were inconclusive. In effect there was again no explicit decision on how long the current payments, and those newly introduced under Agenda 2000, would last. Many farmers believed that the system of payments was fundamentally untenable, and that it was unclear for how long these payments could be sustained or justified. The uncertainty this generates causes considerable anxiety to farmers and their families, and hinders the rational restructuring of farm businesses. This is not only because the debate about payment degressivity over time within the Council of Ministers did not conclude with an explicit decision on the future of direct payments. Farmers were also concerned that eastward enlargement could undermine the sustainability of payments, and that future negotiations in the WTO (including the non-extension of, or a change in, the Peace Clause) might force the EU to reconsider the future of direct payments.

In our original paper we referred to research undertaken at the University of Göttingen that had suggested that the cost to the EU's budget of extending the area and headage payments to the applicant states of Central and Eastern Europe could amount to what we believed to be an unsustainable sum compared with existing expenditure on CAP price and income support (Münch, 2000). In January 2002 the European Commission proposed, in effect, that these payments be phased in over a ten year period in the new Member States, as outlined in Chapter 2. We had also said that there were good reasons for not extending payments introduced in the past in the 'old' European Union to future Member States. Understandably our comments attracted the attention of agricultural economists from the CEECs, and later in this chapter we address some of the concerns of Wilkin (2002). We went on to suggest that it was inconceivable that in the long run there would be two parts of the Union in which different conditions apply regarding direct payments. (Buckwell and Tangermann (1999) discussed these issues in more detail.) It was, and still is, far from clear how these issues will be settled, and this added greatly to the uncertainty about the future of direct payments under the CAP.

Uncertainty of this degree, over a policy instrument that is so central for both farmers and the CAP, is highly detrimental. It has always been one of the most important objectives of the CAP to create stability for farmers. However, the uncertainty regarding the future of direct payments is potentially more detrimental for the viability of farm decision making than market instability and the vagaries of weather can ever be.

Against this background, a first and decisive aim of the bond scheme was to create certainty about the future of direct payments. Under the proposed scheme, an explicit decision would be taken on the duration and future level of payments and on their recipients. Each farmer would then know exactly what payments he or she would be entitled to in the future. The duration of the payments would be limited, and payments would decline over the latter part of their life. However, in exchange for the limitation of payment duration (which is what most farmers expect in any case), there would be an explicit legal commitment that the payments would be made with absolute certainty over a given future period and at a predetermined level. Investment decisions, and decisions regarding the professional future of farmers and their children, could then be taken on firmer ground. Agricultural policy makers could plan future budget allocations and the scope for other measures under the CAP.

A second aim of the bond scheme is to reduce the interference that the existing system of payments has on farmers' decisions. This would result from the 'decoupling' of payments from production. Prior to the Fischler reforms, and in some Member States even now on a continuing basis, farmers needed to plant crops or hold animals in order to receive their area or headage payments. This reduces the economic value of direct payments for the farm. In many cases, some part of a farm's area is planted and some animals are held that generate sales revenues that are less than the costs of production. However, these uneconomic activities are still undertaken because the area or headage payments that they 'earn' are more than sufficient to cover the losses that would otherwise be incurred. Farmers would be better off if they did not have to produce at a loss in order to qualify for direct payments. If payments could be made irrespective of the activities actually undertaken on the farm, farmers could engage in more promising business endeavours rather than making an operating loss just in order to be entitled to payments: a sum of money given to a farmer is always worth more if no strings are attached.

Hence, with a given sum of payments made under the CAP, decoupling these payments from activities on the farm means that all farmers in the EU gain. From the point of view of the overall economy the gains that farmers can make as a result of decoupling are mirrored in a gain in economic efficiency. In other words, if payments are decoupled from farming activities, then gross domestic product in the European Union will grow.

The third reason for suggesting decoupling is that recent CAP reforms should be brought to their logical conclusion. The price cuts were made, in part, to alleviate the surplus problem on the EU's agricultural markets. However, as long as compensation for those price cuts is coupled to the volume of production, the desirable supply response is undermined. This was particularly perverse, for example, on the EU's beef market where the dramatic decline of consumption in response to the BSE crisis resulted in nearly unmanageable surplus problems. It simply did not make sense that the EU destroyed beef to fight the surplus, but at the same time continued to provide production incentives to beef producers by coupling payments to livestock units. During 2001, the evolving saga of foot and mouth disease in the UK vividly illustrated the perverse character of the headage payment system: farmers seeking an exit strategy from livestock production would apparently lose entitlement to headage payments if they failed to restock their farms.

A fourth aim of the bond scheme was to improve the EU's position in the WTO. If payments are truly decoupled, as suggested under the bond scheme, their future would be relatively secure in the WTO, although as we explained in Chapter 2 there are some moves afoot to place limits on overall green box expenditure, and/or narrow the scope of the green box.

A fifth aim was to release budget funds for more constructive uses in rural areas (and potentially the wider EU economy). We broadly endorsed the proposals of Buckwell *et al.* (1997), that the existing CAP should be refashioned into a common agricultural and rural policy for Europe (CARPE), with direct payments *tied* to the provision of environmental and rural services desired by the public, making use of the 'green box' provisions of the WTO *Agreement on Agriculture*[2]. The distinctive feature of our proposal, compared with that of Buckwell *et al.* (1997), is our emphasis on the importance of what they refer to as 'transitional adjustment assistance': this is what we have elaborated as the proposed bond scheme. A second potential call upon budget funds would be to extend the scope of CAP reform to other sectors, notably sugar and milk.

We should make clear that in decoupling compensation payments from future production decisions, bond payments cannot then be made contingent upon any 'cross compliance' requirements relating to land use, animal welfare, or environmental protection. Bond payments would be unconditional. Those farming the land, and tending animals, would have to respect the environmental and other constraints imposed by the state or the EU, or face criminal prosecution. And those providing environmental 'goods,' over and above normal standards of good husbandry, could contract to supply those services in return for CARPE payments which could be financed out of the budget savings made under the bond scheme, under the 'second pillar' of the CAP.

How would a bond scheme operate?

The bond scheme is not a policy that is radically different from the current CAP. Instead, it can be seen as a logical evolution of the existing system of direct payments. They would continue to be paid, albeit in a somewhat revised form. The bond scheme is therefore best described by pointing out the changes to the current regime of payments that would be made. One can well think of a continuum of alternative arrangements for direct payments, with area and headage payments on one end, and a fully developed bond scheme on the other.

In order to explain better how a bond scheme would differ from the pre-Fischler reform payments, we shall describe here six component parts (or steps) of changes that would gradually transform payments into a fully-fledged bond scheme. (In the original proposal we used the term 'successive steps' which led some observers to believe we were proposing a phased introduction of the scheme. This was not our intent. Hence the switch in language to 'component parts'.) The transformation could stop at any of these steps, though we believe that the interests of farmers, EU policy, the overall economy and the international trading regime are met best if the system was fully transformed to a bond scheme.

The individual components, or steps, in which we shall describe the transformation towards a full bond scheme are the following:

1. Decouple crop payments from current land use.
2. Extend this principle to livestock, and decouple livestock payments from the number of animals kept.
3. Decouple payments from land (or the farm) and attach the entitlements to individuals.
4. Limit the duration of payments to, say, ten or twenty years, and (possibly) make payments degressive over time.
5. Definitively fix the future level of payments.
6. Transform payment entitlements into bonds.

As originally proposed in July 2002, the Fischler reforms would have completed steps one and two for most farm products.

Component 1. Decoupling crop payments from current land use

In a first step, the existing arable area payments for crops would be decoupled from production decisions by making future payments to each farm at a flat rate per hectare on the basis of that farm's arable area claims in a recent historical reference period, say a three year period immediately preceding the change in policy. The same principle could apply to other support payments made on an area basis, for example on tobacco. Future payments would then not depend on which crop the farmer actually plants, on the area planted in the future, and not even on whether the land is planted at all. In effect this step would fix definitively an entitlement area for each farm, allowing for unlimited voluntary set-aside or other use because it would not have to be sown with arable crops. It would also eliminate the existing payment differences between different crops.

In Agenda 2000, the EU had already made a significant step in this direction by eliminating the difference between payments on land planted and land set aside, and by aligning oilseed payments with those for cereals under Agenda 2000. There are still higher payments for some specific crops (e.g. protein crops and durum wheat). Such differences between crops would disappear, and in the future, only a flat rate payment per hectare on which area payment claims had been made in an historic reference period would be made.

With this step, crop payments would be completely decoupled from future production decisions and land use. Hence most of the advantages of decoupling described above would already be reaped for crop payments. Payments would, however, continue to be made to the farmer holding the right to use the land that formed the entitlement area of the farm concerned during the reference period. This, essentially, was what the Commission proposed for the mid-term review.

Component 2. Decoupling livestock payments from the number of animals kept

Decoupling of livestock payments from the number of animals held had already (implicitly) existed in those instances when farms hold *more* animals per farm or per hectare than the limits up to which payments are currently made. However, if farms were to reduce livestock numbers below these thresholds, then payments would be linked directly to livestock numbers. Thus, general decoupling of all livestock payments requires a slightly different approach. The number of livestock units (cattle, goats and sheep) on which future payments are based could be the number of livestock units for which the farm claimed payments in a recent historical reference period, say the average of the three most recent years before the policy change. As in the case of crops, that entitlement would be held constant, irrespective of the actual number of livestock units that the farmer concerned kept in the future. Farmers could then be allowed to attach that payment entitlement to a land area of their choice and, if the farm was subsequently broken-up, future payments would continue to be made to the component parts. Again, this is what the Commission proposed for the mid-term review, but the Council enacted a more complicated package.

Component 3. Decoupling payments from land (or the farm) and attaching the entitlements to individuals

Binding payment entitlements to land, as suggested in 1 and 2, has the advantage of keeping them as similar to the current regime as possible. But it does decouple payments from production decisions and therefore most of the significant benefits resulting from decoupling are reaped. In that sense, components 1 and 2 are a most important part of transforming current payments. However, if payments remain tied to land, then their value will continue to be reflected in land prices, for both purchase and rental. The value of land continues to be distorted, reflecting not only the market value of the products generated on the land, but also the effects of government policies. For structural change in agriculture this is a significant disadvantage, as farmers who want to expand then continue to find it difficult to finance the acquisition of larger areas for their farms. At the same time, farmers entitled to payments cannot disentangle their payment rights from their ownership of land, and their flexibility is hence reduced.

For these reasons, a sensible third component is to de-link payments from land. As the justification of the payments is to help farmers adjust to the reformed

CAP, the most appropriate subject for future entitlement to payments is the farmer as a person (or legal entity, such as a limited liability company). As far as administrative modalities are concerned, this is easily achieved by detaching the payment entitlements from land, as described in 1 and 2, and transferring them over to the farmer holding the right to farm that land at the moment this step is introduced into the CAP (or, alternatively, holding the right to farm that land in a recent historical reference period). The farmer would then receive a document entitling him (or her) to receive the future payments that would otherwise have been made to his land or farm. He would keep this document, and would continue to receive the payments, irrespective of what he does with his land. Even if he were to sell or to rent the land, the payment entitlement would remain with him, rather than moving with the land. If the farmer should die before the final direct payment had been made, his or her heirs would succeed to the entitlement.

A difficult question, of course, is how different categories of rights to farm land would be treated. In particular, where land is farmed on a rental basis at the time when payment entitlements are separated from the land and handed over to persons, would the entitlement be given to the tenant or to the landlord? We return to this question below and in Chapter 7.

Component 4. Limiting the duration of payments (and possibly making payments degressive over time)

In this, and the following component of transformation, certainty is generated over the future stream of payments. As argued above, the current uncertainty about the future of the payments under the CAP is an untenable situation. Most farmers realise that the current payments will not be made forever, but have no idea how long they will last. In this situation, it would help if an explicit decision was taken on the duration of the payments that are currently made. Politically this will not be easy, and it may appear to be more attractive to leave that decision to future governments. However, farmers have the right to know about future policies on which their business and family decisions crucially depend, and policy makers hence have a moral obligation to let them know.

If this is not done, the same fundamental problems will arise as in the past under the policy of seemingly open-ended price support. When that policy was ended abruptly, farmers were caught by surprise and therefore had an undeniable right to compensation/adjustment help. The same must not happen again. A situation must be avoided in which farmers, told without warning that payments will end, can claim the right to receive adjustment aids because old payment schemes are eliminated. The only way to avoid this is to take an early and explicit decision on the duration of future payments.

What that duration should be is essentially a political question, on which economists can only make suggestions. The remaining duration of payments should be long enough to allow farmers to make all necessary adjustments so as to be able to live with the policy change that was made when the payments were introduced in the first place. It should, on the other hand, not be so long that farmers have no incentive to make the adjustments required by the reform of past price policies. From a purely political point of view, the period should not be so short that it is difficult to believe that payments will really be ended then, but should also not be so long that its end appears to be unrealistically far away. In our 2000 Working Paper we suggested that a period anywhere between ten and twenty years would probably meet these criteria: 'Perhaps 15 years of further payments after the transformation is made is a reasonable duration for future payments' (Swinbank and Tangermann, 2000). In the questionnaire sent to farmers, reported in Chapter 8, we suggested that the transformed payments should continue over a 15 year period, which gave rise to some interesting responses. We want to emphasise, however, that it is not the precise duration of future payments that is so important (to us), but rather that a fixed deadline is firmly agreed and publicised.

One other parameter for future payments, closely related to the duration of payments, is a possible decline of payments over time. Again it is a largely political decision whether payments should be made at their full initial level until the last year of the scheme, or whether they should begin to decline at some stage. Something can probably be said for a degressive schedule, because it makes farmers better aware of the need for adjustment, and also because most farmers will be adjusting their farming practices over time, thereby reducing the need for further payments. This may also have been a reason considered by the Council of Ministers when degressivity of payments was discussed in the debate about Agenda 2000. If degressivity is considered sensible, and if a 15 year payment profile should be adopted, then the schedule could be arranged such that, for example, payments are made in full until year 11, and then decline by 20% per year, reaching zero after year 15.

Component 5. Fixing definitively the future level of payments

Full certainty about future payments requires not only that their duration, but also their level, during the remaining lifetime, is decided explicitly and irrevocably. Again this is a significant problem with current arrangements, where farmers begin to feel uncertain over the levels of future payments. What that future level should be is another purely political decision. However, a simple approach would be to suggest that all payments are kept at their current (or already fixed future) level until the year before degressivity cuts in.

As well as the certainty this implies for farmers, it removes from agriculture ministers the annual temptation to renegotiate payment levels. In effect, provided the legislation is properly crafted, in one decision ministers set the future level of payments over the lifetime of the scheme. Thus if a future Agriculture Council wanted to make additional payments to farmers, this would require new legislation which would be harder to introduce.

Component 6. Transforming payment entitlements into bonds

If all component parts, including 5, are made then each holder of payment rights has a document that effectively entitles him or her to receive a predetermined flow of annual payments, over a predetermined future period. A last and useful step would then be to guarantee that stream of payments to the holder of that document, whomsoever he or she may be. This would allow the original recipient of that document (determined in 3 above) to sell the document on the financial market at any time, thereby effectively converting the guaranteed stream of annual payments into a sum of money that can be used, for example, to make investments on the farm to prepare it better for the new market conditions, or to create a new earnings opportunity outside agriculture.

In this step, the government document guaranteeing the future payments would acquire the characteristics of a government bond with interest payments only, without a principal component[3]. The capital value of that bond would be determined on the market, depending mainly on interest rates (current and expected).

It should be emphasised again that the step-by-step description of the bond scheme given here is mainly to help an understanding of how a bond scheme would relate to, and differ from, the current regime of direct payments. Of course all components described here separately could (and we would argue should) be taken at the same time, thereby transforming the current payments into a bond scheme in one single step. However, it should also be clear that significant improvements to the current regime could also be made even if not all steps are taken. Hence, the bond scheme described here should be seen as a desirable form of complete decoupling of payments, but not as the only form of decoupling that would make sense.

It should, though, also be noted that there is a given logical relationship between these individual steps, which build on each other and can therefore be taken only in the sequence presented here. The only exception is step 2 (decoupling of livestock payments), which can, but does not necessarily have to, be part of any subsequent step. Most importantly, step 3 (moving payments from a land base to a person base) can only be taken if payments have been completely decoupled from production, and step 6 (transforming payments into bonds) requires that both the duration and the level of payments have been irrevocably determined in advance.

What are the benefits of bonds?

A bond scheme introduced in this way would have all the advantages of decoupling described above and some other benefits, all summarised in Table 4.1. With a given level of payments, farm incomes would rise, and the overall EU economy would be better off. Farmers would have more flexibility to decide on land use and indeed on their future professional activities. Quasi-mandatory set-aside would go, and government interference with farmers' decisions on land use would end.

Table 4.1. Summary of the benefits of a bond scheme

Component/Step	Benefits
1 and 2: Decouple payments from production	Allows farmers to make more productive use of their resources. Alternative uses of farm land (e.g. the purchase for use as bird sanctuaries) would become more feasible. The IACS system, and related administrative controls, can be dismantled unless required for traceability of livestock. Payments switch from the WTO's contentious blue box into the internationally accepted green box.
3: Decouple payments from land (or the farm) and attach entitlements to individuals	Land prices are no longer distorted by the capitalisation of expected future payments, further facilitating farm restructuring. New entrants into agriculture no longer have the expectation of receiving payments.
4: The period over which future payments will be made is fixed	Re-inserts a level of certainty into policy, enabling better-informed farm investment decisions to be taken.
5: The level of future payments is fixed irrevocably	Reinforces the level of certainty in the industry. Enables all aspects of the existing legislation to be repealed, making it more difficult for future agriculture ministers to amend payment levels.
6: Introduction of bonds, and the full transferability of payment entitlement	Locks-in policy reform, as payments cannot be altered without impacting upon the wealth of bondholders who are no longer the original farmer recipients. Enables the original recipients to sell their bonds on the stock market, releasing funds for alternative uses.

Another significant advantage is that the huge administrative effort which is currently made in controlling farmers' behaviour (hectares planted to eligible crops, livestock units kept) and making the payments on that basis would no longer be necessary. We originally claimed that the whole Integrated Administration and Control System (IACS) could simply go, but IACS in fact has a subsidiary use in that it provides for the traceability of livestock introduced in the wake of the BSE scares. Payments could be automatically made to bond holders, possibly through banks, in the same way that interest payments are made on government bonds.

For the future policy process, a significant advantage is that the policy change made in this way is locked in. Furthermore, as the bonds can be held by anybody, future payments would in effect no longer be a matter of agricultural policy, but rather a predetermined matter of a relatively simple financial nature.

How does the bond scheme differ from the Fischler reforms?

In the mid-term review, as outlined in Chapter 2, the European Commission proposed a decoupling of direct aids and the establishment of a Single Payment Scheme (Commission, 2002, 2003), reflecting components 1 and 2 of the proposed bond scheme (see Table 4.2). Thus the advantages identified in Table 4.1, associated with adoption of components 1 and 2 of the bond scheme, will be reaped by Mr Fischler's reform. The Single Payment Scheme, however, did not adopt component 3 of the bond scheme, totally decoupling the payment from the farm. Instead a cross-compliance element has been introduced to enforce 'good farming practices', buttressed by farm audits, and with a requirement for 'environmental set-aside'. A Reading student, in an examination answer, referred to 'double decoupling' in the bond scheme in contrast to the partial decoupling in the Single Payment Scheme.

Table 4.2. Comparison between the bond scheme and the Single Payment Scheme of the Fischler reforms

Bond Scheme	Single Payment Scheme
1 and 2: Decouple from production	Only where so-called full decoupling applies
3: Decouple from land	No, the Single Payment Scheme is tied to the farm (the historic payment scheme) or land (regionalised scheme). Payments are conditional on cross-compliance and 'environmental' set-aside
4 and 5: Fix, and phase out	Paid for an unspecified period. For farms receiving in excess of €5,000 per annum, payments subject to modulation reaching 5%, with the threat of future reductions under the new Financial Discipline
6: Transferable bond	No

Unlike the bond scheme, under which the profile of payments would be irrevocably fixed, no such certainty was built into the Single Payment Scheme introduced by the Fischler reforms. The scheme is in place for an indefinite period, giving little certainty to producers as its terms could be changed at any time. Producers could reasonably suspect that, at some future date, the rate of modulation could be increased beyond 5%. Indeed, the regulation introduces a new *Financial Discipline* under which future payments can be reduced if budget forecasts suggest that the annual expenditure ceiling, less €300 million, is exceeded (Council, 2003). Thus, although a welcome step in the right direction, it clearly fell short of our proposed bond scheme. Nonetheless, now adopted, it will be that much easier in the next reform to adopt a bond scheme in its entirety.

What problems could arise?

Our initial reflections, and discussions with policy makers and stakeholders, raised a number of issues and concerns that helped shape the questionnaire we used to survey farmers' intentions, as reported in Chapter 8. In this section, we address some of the more theoretical concerns encountered in our deliberations. In Chapter 7 we outline some implementation issues, and in Chapter 8 we report on what farmers in Germany, Portugal and the UK said they would do if a bond scheme was introduced.

The issues addressed in this section are:

1. Will too much land fall idle?
2. Will too many farmers leave agriculture?
3. Should landlords or tenants be the recipients?
4. Should there be modulation (lower per unit payments for larger farms)?
5. What reference period should be used as a base?
6. Should there be provisions for responding to unexpected price changes?
7. Can conversion into bonds be made voluntary?

Will too much land fall idle?

When farmers receive payments irrespective of how they use their land, marginal land may well drop out of production. There could be fears that such idling of land may be concentrated in some regions, with negative implications for the countryside and rural development. However, we felt that the extent to which this might happen should not be overestimated. Under the old regime for the *grandes cultures*, voluntary set-aside of land was already possible to some extent, and the regions where this happens are also most likely to be the ones where land would be idled under a regime of decoupled payments. So far it appears that voluntary set-aside has not become a major issue (otherwise it would already have been eliminated by policy makers)[4]. However, under a regime of fully decoupled payments there would not be an upper limit to idling, and whole farms could potentially be closed down. Hence there might be fears that the area affected could grow. Such fears should, though, not be exaggerated because there are various forms of extensive land use that might spread if land has no longer to be planted to the *grandes cultures* in order to receive payments. With decoupling of livestock payments in the EU, idling might also spread to areas now used for extensive goat, sheep and cattle operations, currently pursued in order to receive livestock payments. The response to our survey, detailed in Chapter 8, gives some idea of how farmers might react.

However, even if there should be the expectation that too much land might fall idle in some regions, this is not a good reason to argue against decoupling, for two reasons. First, this would be a problem concentrated in certain regions. It does not make good economic sense to shape the regime of payments for the whole territory of the EU such that the problems of only some regions are solved. A much better approach, and less costly to the overall EU economy, is to run specific policies in those regions where such problems actually arise. Second, the current regime is unfair to farmers farming the land concerned, in the sense that it requires them to plant or graze that land even if they make a loss in market terms (i.e. before receipt of the area or headage payments) only in order to receive the payments. In effect this means that these farmers are required to provide a free service to society, by losing part of the payments which were established to compensate them for past price reductions. For both reasons decoupling should be allowed to go ahead, possibly accompanied by selected regional programmes that provide for continued use of land in a form desired by society. If the extra payments possibly needed for this purpose are considered to be too much of a budget burden, then the decision could be taken to reduce payments to all EU farmers on a proportional basis and use the savings for financing the specific regional programmes. This approach would be fairer to the farmers in the regions concerned than the current regime.

Some commentators have debated the potential impact on the stewardship of the land if CAP price support were to be reduced. However, under the proposal presented here the environmental impact would be minimised as the intensity of land use would not change. Area payments are made on the area planted and not on actual yields. Hence area payments are already decoupled from yields. This would not change under a bond scheme.

A more potent problem is the risk that land might be diverted from arable crops at the intensive rather than the extensive margin, with farmers switching to the production of other CAP supported commodities. However, milk and sugar production is controlled by quota, and so an increase in the output of these commodities is precluded. Again, the questionnaire - discussed in Chapter 8 - was designed to gauge farmers' likely response.

Will too many farmers leave agriculture?

Decoupling of payments from land use does not reduce farm revenues; on the contrary it allows farmers to use their resources more economically and therefore leads, for a given level of payments, to higher farm incomes. From this perspective there will be not more, but instead less, pressure for farmers to leave the land. However, as area no longer has to be planted, and animals no longer have to be kept in order to receive payments, the amount of labour input required in agriculture may decline if and where land use and livestock production decrease as a result of decoupling. At the same time, with full decoupling, farmers no longer have to remain in agriculture just to receive payments, and some may indeed decide to leave. This should be seen as a positive development as it provides some farmers with the opportunity of looking for better paid jobs, without any immediate income pressure necessitating such a move. This effect is strengthened even more if payments are converted into bonds, thereby allowing farmers to invest in new opportunities outside agriculture if they want. To the extent that farmers respond to these new options, structural change in agriculture is facilitated and the competitiveness of EU agriculture is strengthened.

Should landlords or tenants be the recipients?

Where farms are owner-operated there is no debate about who should be the recipient of bonds. It would be the farmer owning and managing the farm. However, potentially rather contentious issues arise in cases of share-cropping, seasonal grazing licenses, conacre, landlord-tenant systems, or partnership arrangements where the partners undertake different functions. Who should receive the bonds in such cases? In order to find an appropriate answer to this question one has to consider the effects on income distribution between the different agents that arise under the current regime of payments. Current payments are generally made to the person holding the right to use the land on the respective farm. Thus, on rented farms payments go to the tenant rather than the landlord. However, where land is rented or share-cropped, rents or share-cropping arrangements tend to reflect the value of the payments. In other words, payments tend in effect to flow to the owner of the land.

Based on this consideration, one could argue that the bonds should go to the landowners. But then a corresponding adjustment would need to be made to the rental price (or share-cropping arrangement) in the respective contracts. Since legal conditions differ from country to country, and often from case to case, this matter is probably best left to the Member States, which would have to enact appropriate legislation. In a way this can be compared with the arrangements that had to be found when milk quotas were issued in 1984. In that case, the issue of ownership allocation also had to be resolved where dairy farms were not operated by the landowners, with similar economic implications. The legal treatment of such cases was left to the Member States.

Should there be modulation (lower per unit payments for larger farms)?

The issue of modulation of payments (i.e. lower per unit payments for larger farms) is not specific to a bond scheme. It has arisen several times in the past under the old payment regime. In essence this is again a political issue. If felt politically appropriate, the amount of bonds issued to individual farms could be modulated. However, it is only at that point in time when modulation could be introduced. Once bonds have been issued they can be sold to anybody, and modulation at a later point in time would no longer be appropriate or possible.

What reference period should be used as a base?

The reference period to be chosen for allocating the amount of bonds per farm should, as in similar cases in the past (e.g. introduction of milk quotas), be a recent historical reference period, say the average of the most recent three years. However, because of the large economic significance of this policy change it would be particularly important to avoid announcement effects (i.e. farmers taking decisions just in order to maximise future payments). Hence the reference period should be chosen sufficiently far in the past, so as not to be unduly influenced by decisions that may have been triggered at a time when the political debate about a possible introduction of a bond scheme began. The fact that the existing direct payments are in any event constrained by entitlement criteria does, however, limit the potential impact of any announcement effects.

Should there be provisions for responding to unexpected price changes?

In the history of agricultural policy there are several cases when decisions that were claimed to have been made on a permanent basis were later revised because circumstances had changed. An obvious example is the large bail-out payments that were made in recent years to US farmers, though the payments introduced under the 1996 FAIR Act were supposed to have been fixed irrespective of future developments (Ayer and Swinbank, 2002). These extra payments in the US were made in response to lower-than-expected market prices. Under a bond scheme this must not, and cannot, happen because the owners of the bonds will not necessarily be the farmers affected by contemporaneous developments on agricultural markets. For the agricultural policy process this is a big advantage. Farmers, though, may see this differently.

However, it should be abundantly clear from the start that the introduction of a bond scheme is a final act as far as the policy changes covered by the bonds are concerned. This is not to say that all market and price policies necessarily have to disappear. As with the current area and headage payment regime there can still be market policies (e.g. in the form of intervention prices) that provide a safety net for prices. It would be for such market policies to provide for security against price drops, if politically it is felt that this is necessary.

Can conversion into bonds be made voluntary?

In past discussions about bonds as an instrument of farm policy reform it has been suggested that bonds could be introduced on a voluntary basis (Poole, 1993). Individual farmers, it was suggested, could be allowed to choose between bonds and a continuation of the current payment regime. We suggest that this should not be a feature of a bond scheme, for several reasons. With voluntary bonds, the savings in administration could not be made; distortions of land prices would be maintained (and would in fact be exacerbated); the effect of locking in the policy change would disappear; and payments that continue to be made under current arrangements could not go into the WTO's green box. For such reasons we suggest that bonds are introduced for all farmers.

Reaction to the proposed bond scheme

Concern has been focussed on the unfavourable impact of the proposal on the accession states of Central and Eastern Europe, particularly Poland (Wilkin, 2002)[5]. He saw it as a manoeuvre to allow the EU to avoid burdening its budget with the cost of extending existing headage and area payments to the new Member States. We used the argument, less and less accepted by other agricultural economists he claimed, that these payments compensated for the loss of farm revenues brought about by the MacSharry reforms, and as the new Member States did not experience these price cuts there was no reason to provide for compensation. But under the bond scheme there would still be a substantial flow of funds into the farm sector in the western half of the EU, and this would be discriminatory and undermine the principle of equal competitive conditions for all EU members. There is greater need for adjustment assistance in the new Member States, he argues, than in the EU-15, although the latter also need support in adjusting to a world of freer trade. The proposal as it stood amounted to a 'veiled' discriminatory treatment of Central European farmers. In the event, at the Copenhagen Summit of December 2002, the EU decided to extend direct payments to the New Member States as outlined in Chapter 2.

Ingersent (2002), in criticising our approach, expressed concern about the stability implications of our proposal, and favoured deficiency payments instead. We felt that he took our proposal out of context (Swinbank and Tangermann, 2002). We had advocated that the existing area and headage payments under the CAP be replaced by a fully decoupled transferable bond, and had suggested that this should be 'embedded in a comprehensive reform of the CAP to reflect better the public's wider concerns about the environment and rural development'. In our necessarily short article for *EuroChoices* we did not discuss details of that comprehensive reform. However, we had broadly endorsed the proposals of Buckwell *et al.* (1997) 'that the existing CAP should be refashioned into a common agricultural and rural policy for Europe (CARPE)'. In the Working Paper on which the *EuroChoices* article was based, we had stated that the introduction of a bond scheme to replace existing direct payments did not mean that 'all market and price policies necessarily have to disappear. As with the current area and headage payment regime there can still be market policies (e.g. in the form of intervention prices) that provide a safety net for prices. It would be for such policies to provide for security against price drops, if politically it is felt that this is necessary' (Swinbank and Tangermann, 2000). It is true that one of the co-authors had elsewhere advocated more sweeping reform. This did not, however, deny the notion that EU and national governments 'might facilitate the introduction of risk management strategies for the farm sector to cope with harvest and price uncertainty, such as futures markets, forward contracting, crop insurance, and the averaging over several years of income for taxation purposes' (Beard and Swinbank, 2001).

In short, we had not proposed that bonds be the only remaining instrument of the CAP. What we did propose is that the implementation of what has become the central instrument of the CAP, i.e. direct payments, be changed. We would indeed envisage other instruments in the future that help farmers cope with market instability, though these instruments should be designed such that they interfere as little as possible with market forces.

The implication of Ingersent's statement that, 'the standard (*a priori*) economic case for stabilising producer prices appears to be in danger of being forgotten' (Ingersent, 2002) seems to be that area and headage payments do provide stability, which would then be lost if they were converted into bonds. However, this is not really true. Current payments do not respond to fluctuations in yields and/or prices, and hence do not stabilise any variations in market returns. As a logical implication, decoupling of current payments and their conversion into bonds does not as such result in less stability.

What could be argued, though, is that eventual elimination of direct payments (in their current form or in the form of bonds) reduces stability. Indeed, payments provide a secure *revenue* flow for farm businesses, in addition to fluctuating revenues from market sales. But over time these payments have been largely capitalised into the cost structure of the farm business (in particular into land prices), and the residual income flow is still subject to the shocks of yield and price changes. Thus we believed that there was a very tenuous link between the goal of stabilising producer returns and an indefinite continuation of area and headage payments. Moreover, compared to the political uncertainties inherent in the existing schemes, our proposed bond scheme would allow for greater certainty in the flow of compensation payments over a defined period, allowing the farm sector to adjust to new economic realities. In other words, Ingersent forgets about policy uncertainty and considers only market fluctuations.

Ingersent's idea that deficiency payments might stabilise, but not raise, producer prices is novel (Ingersent, 2002). Our understanding of a deficiency payments system is that it is asymmetrical, i.e. making-up producer prices when they fall below a guaranteed level, but allowing them to rise above the guaranteed level in other market circumstances. The same would be true for the PEG ('production entitlement guarantee') proposal cited by Ingersent (2002) which, for example, Blandford *et al.* (1989) discussed. Of course, in theory, deficiency payments (and a PEG regime) could be implemented such that subsidies are paid when market prices are low, while taxes are collected when market prices are high, thus achieving (price) stability without, on average, providing support. Deficiency payments, though, do not eliminate revenue fluctuations originating from yield shocks. But we do not believe that this is a politically realistic option. Nor did Ingersent (2002) suggest that approach.

Moreover, deficiency payments do not address the fundamental problems that we had identified as inherent in existing policy: for example, the economic inefficiencies in the partial coupling of payments to production; the difficulties of negotiating the retention of blue box payments under a new WTO Agreement; and the cost of extending these payments to acceding states in Central and Eastern Europe. In that regard, a PEG regime would do only marginally better than deficiency payments. Also, Ingersent appeared to ignore the fact that a policy of stabilising domestic prices exports price instability to world markets, thereby creating part of the problem that the policy is supposed to solve.

Henning (2002) and Mortensen (2002) were concerned about the political acceptability of our approach. Mortensen points out that, compared to 1992 when there was debate over the scope of a bond scheme (the payment base and the level of payment), the present proposal (existing area and headage payments) is well defined. Henning argues that 'given the specific logic of the political economy of agricultural policy, it still is today, as it was in the past, politically infeasible'. In Chapter 6, Daugbjerg revisits the 1991/92 debate over the Danish government's espousal of a bond scheme, and examines the political feasibility of a bond scheme proposal today.

The Producer Bond, advocated by the Tenant Farmers' Association (2002) in a submission to the Select Committee on Environment, Food and Rural Affairs of Britain's House of Commons, was a complex device that has some resemblance to our proposal. As with our bond scheme, the *total* annual payment receivable by a farmer would be determined by reference to an historic entitlement to direct payments. But there would be at least two components to their scheme: one element would be a guaranteed annual sum (perhaps index linked), decoupled from production, that would be paid for a guaranteed period, and through the financial markets this annual stream of payments could be capitalised. In response to questioning by the Select Committee, the Chief Executive of the Tenant Farmers' Association (2002) explained that they did not expect that these payments would terminate once the guaranteed period had lapsed; but instead that they could be renewed for another guaranteed period. A second element of the overall package would be 'annual payments to producers in return for an annual contract covering environmental, rural development and animal welfare issues'. Again it is suggested that farmers would have the right to 'assign or capitalise these payments', though it is not clear how cross-compliance can be imposed on a particular farm if the entitlement to payment is owned by another person (or company) that has no managerial control over the farm. Thus, to return to the triangle depicted in Fig. 1.1, the Tenant Farmers' Association (2002) envisages a policy mechanism that continues to provide long-term income support, and helps secure some of the multifunctional outputs of the agriculture sector; it is not a compensation package, occupying the apex of the triangle, as we envisaged in our proposal.

What role could bonds play in future reforms?

Up to this point, the approach of decoupling payments, possibly all the way to a bond scheme, has been described as a strategy to improve on the functioning of the current regime of payments. Thus, from one perspective, it could be argued that the proposal for the decoupling of existing payments and their replacement by a bond scheme is not a fundamental reform of the current CAP. Reform decisions have already been taken, in 1992 (MacSharry), 1999 (Agenda 2000) and 2003 (Fischler), and decoupling/bonds would only make one element of those earlier reforms, the direct payments, more certain and effective.

Writing in 2000, we said that by about 2002 or 2003, with the prospect of eastern enlargement looming ever closer, and the WTO negotiations on agriculture drawing to a close, a further reform of the CAP would be needed (Swinbank and Tangermann, 2000). Although from a 'northern' perspective a 'stand-alone' bond scheme to replace the existing area and headage payments might seem to be a logical outcome of this review, Ministers from Southern Member States, and members of the European Parliament, might take a more holistic view. A bond scheme which offered some EU farmers a secure pattern of payments over a 10 or 15 year time horizon, whilst the continuation of support for other producers remained uncertain, set in the context of a CAP reform which did little to develop the second pillar of the CAP, might well appear unattractive. Thus the 2003/04 reform would need to be set in this wider context for it to command majority support. In the event, the European Commission's proposals for the mid-term review can be seen to be positioned with this conflux of interests in mind.

The Fischler reform is unlikely to be the last step in CAP reform. Support prices for milk may have to be reduced even further on the way towards eliminating milk quotas. Reform of the sugar regime may one day appear feasible. The market regime for horticultural products, including the specific border measures under that regime, may come up for review. In some or all of these cases the need may be felt to make (additional) direct payments in order to help farmers tide out the necessary adjustments. All these future payments could, and we believe should, be made as decoupled payments or even in the form of bonds with immediate effect. There would then be bonds of different future durations, and with different payment levels. However, that is no problem, as the financial markets can easily handle different instruments of this type in parallel.

We would go one step further and argue that CAP reform should indeed continue, and that a bond scheme is a good basis for such further reform, though much would be gained already if only the current payments were transformed into bonds. We suggest that further steps towards CAP reform, each with payments in the form of bonds, should be taken in the near future.

A staggering of these sector-by-sector reforms over several years would help to distribute the budget burden resulting from payments over several years, and some of the payments for the later-reformed product sectors would then fall in a period when payments under the 'old' reforms had already begun to decline.

At the same time as aggregate expenditure on direct payments (possibly in bond form) related to 'old' and 'new' reforms of market and price policies declines over time, financial scope is created which allows the EU to engage increasingly in environmental and rural development policies (the 'second pillar' of the CAP).

Figure 4.1 illustrates the scenario we have in mind and is illustrative only, and not drawn to scale. The existing annual budgetary expenditure on direct payments is represented by A. Under existing legislation, this expenditure is set to continue largely unabated. We have argued that this is untenable.

An alternative scenario would involve a policy change with immediate effect. This would switch the existing direct payments into annual payments to bond holders (represented by B in Fig. 4.1). Initially, EU budget expenditure would be unchanged, but after a specified period annual payments would be reduced and eventually phased out. This would release budget funds for alternative uses, including the funding of bond schemes to facilitate reform in other sectors (C), and to provide enhanced funding for environmental and rural development programmes tied to specified deliverables in conformity with the WTO's green box (D in Fig. 4.1).

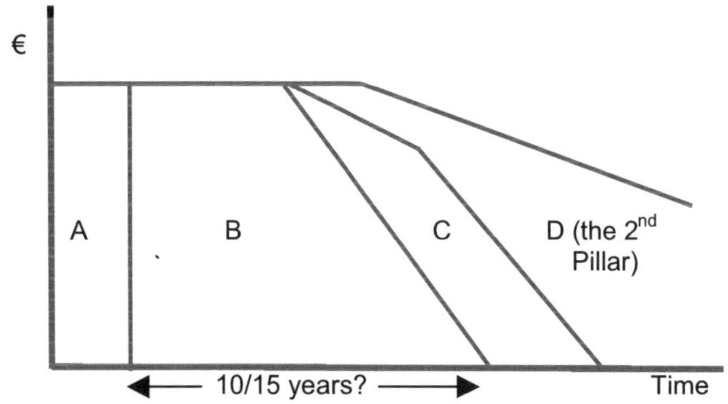

Fig. 4.1. Possible expenditure on the bond scheme and wider CAP reform

Under this scenario, the size and form of C and D need to be negotiated. Arguably, because of the regionally oriented nature of environmental and rural development policies and under the subsidiarity principle, some or all of D could be funded directly by national exchequers, rather than via the EU budget; but this consideration lies beyond the remit of this research project.

Conclusions

In this chapter we have not, in the first instance, advocated fundamental reform of the CAP, but rather a further evolution of the MacSharry reforms of 1992, and the Fischler reforms of 2003. We have argued that there is considerable uncertainty about the future of direct payments, and have suggested a mechanism under which certainty about their future could be re-inserted into the policy. Farmers would have an assured profile of payments, allowing them a period to adjust to the policy reforms inaugurated in 1992. Another key element of the proposal is the decoupling of payments from production, removing from farmers the constraints on land use and livestock production that the present system implies, and thereby enabling them to produce more efficiently.

We have advocated a six-point evolution of the payments system, which in its entirety would amount to a system of fully-transferable bonds. Although many of the advantages of decoupling can be reaped by adopting only one or two steps in this transition, the full benefits - to farmers, and the wider economy - are dependent upon the whole. The full bond scheme, in addition to removing the shackles which force farmers to plant some hectares or keep some animals uneconomically in order to receive payments, would also remove much of the IACS administrative controls which are burdensome to taxpayers and farmers alike, and allow for more economic use of Europe's farmland, contributing towards improved international competitiveness.

Over time budget funds would be released for other, more constructive, uses. Thus a bond scheme could also be deployed to facilitate further reform of the CAP, particularly for milk, sugar and Mediterranean crops. This will undoubtedly prove necessary as a result, *inter alia*, of eastern enlargement and the ongoing process of international agricultural policy reform under the auspices of the WTO. It would extend to these producers the enhanced security of a guaranteed flow of compensation/adjustment payments that a bond scheme can offer. Furthermore, although the prime purpose of the present chapter was to argue the merits of a fully decoupled system of compensation (or adjustment) payments, as exemplified by the proposed bond scheme, we would see this embedded in a comprehensive reform of the CAP to better reflect the public's wider concerns about the environment and rural development.

Notes

[1] Net value added at factor cost for agriculture in the EU-15 in 1998 amounted to 107.4 billion ecu. Expenditure on 'compensatory aid - reform' of 21.8 billion ecu is equivalent to 20.2% of that net value added at factor cost.

[2] We have used the word 'tied' in this sentence, rather than 'coupled', to emphasise the fact that our proposal involves the decoupling of direct payments

from production, with CARPE payments explicitly tied to the provision of environmental and rural services.

[3] Adrian Baird, in a personal communication, noted: 'The document of entitlement could, perhaps, be better described as having acquired the characteristic of a bundle of stripped gilts'. In Chapter 7 we discuss some of the tax complications that could arise in the UK, depending upon the particular form the entitlement took. We also are grateful to Peter L. Arcus for pointing out that, strictly speaking, the annual payments under our proposal are analogous to the interest strips, rather than the stripped bond which repays a capital sum on maturity, that the financial markets create when they un-bundle the component parts of conventional government bonds.

[4] In 1996/97, area idled in the EU-15 under mandatory set-aside (at the mandatory rate of 10%) was 3.7 million hectares. Voluntary set-aside amounted to another 1.9 million hectares, i.e. roughly 5% of the base area for *grandes cultures*. Source: www.europa.eu.int/comm/dg06/publi/cap2000/cereals/cerealde /cerealde.pdg

[5] We are grateful to Alexandra Trzeciak-Duval for help in reading Wilkin's paper; this paragraph borrows freely from Wilkin's text.

References

Ayer, H. and Swinbank, A. (2002) The 2002 U.S. Farm Bill: what's in it for CAP and WTO reform? *EuroChoices* 1(3), 26-32.

Beard, N. and Swinbank, A. (2001) Decoupled payments to facilitate CAP reform. *Food Policy* 26(2), 121-145.

Blandford, D., de Gorter, H. and Harvey, D.R. (1989) Farm income support with minimal trade distortions. *Food Policy* 14, 268-273.

Buckwell, A., Blom, J., Commins, P., Hervieu, B., Hofreither, M., von Meyer, H., Rabinowicz, E., Sotte, F. and Sumpsi Vîna, J.M. (1997) Towards a common agricultural and rural policy for Europe. *European Economy* Reports and Studies No 5. CEC, Luxembourg.

Buckwell, A. and Tangermann, S. (1999) Agricultural policy issues of European integration: the future of direct payments in the context of Eastern enlargement and the WTO. *MOCT-MOST: Economic Policy in Transitional Economies* 9, 229-254.

Commission of the European Communities (2002) *Communication from the Commission to the Council and the European Parliament. Mid-term Review of the Common Agricultural Policy*. COM(2002)394. CEC, Brussels.

Commission of the European Communities (2003) *Proposal for a Council Regulation Establishing Common Rules for Direct Support Under the Common Agricultural Policy and Support Schemes for Producers of Certain Crops...(A Long-term Perspective for Sustainable Agriculture)*. COM(2003)23. CEC, Brussels.

Council of the European Union (2003) Council Regulation (EC) No 1782/2003 of 29 September 2003 establishing common rules for direct support schemes for farmers. *Official Journal of the European Communities* L270, 1-69.

Henning, C. (2002) On the political economy of economically efficient compensation schemes. *EuroChoices* 1(2), 44.

Ingersent, K. (2002) CAP bonds: a sceptical view. Counterpoint. *EuroChoices* 1(2), 30-32.

Mortensen, J.R. (2002) Is a CAP bond scheme more feasible today than it was ten years ago? *EuroChoices* 1(2), 43.

Münch, W. (2000) Effects of CEEC-EU accession on agricultural markets in the CEEC and on government expenditure. In: Tangermann, S. and Banse, M. (eds) *Central and Eastern European Agriculture in an Expanding European Union*. CAB International, Wallingford, UK.

Poole, P. (1993) *Reform of the Common Agricultural Policy: the Use of Exit Bonds*. European Policy Forum, London.

Swinbank, A. and Tangermann, S. (2000) *A Proposal for the Future of Direct Payments Under the CAP*. Working Paper No 1 of the Bond Scheme Project. Centre for Agricultural Strategy, The University of Reading, UK.

Swinbank, A. and Tangermann, S. (2001) The future of direct payments under the CAP: a proposal. *EuroChoices* Premier Issue, Spring, 28-29, 32-34.

Swinbank, A. and Tangermann, S. (2002) CAP bonds: a sceptical view. Reply. *EuroChoices* 1(2), 33-34.

Tenant Farmers' Association (2002) Memorandum and Minutes of Evidence in House of Commons Select Committee on Environment, Food and Rural Affairs. *Minutes of Evidence, Wednesday 8 May 2002*, HC-550-x. The Stationery Office, London.

Wilkin, J. (2002) Komentarz. *Wies I Rolnictwo* 114, 106-107.

Chapter Five:

From CAP to CARPE: Embedding the Bond Scheme Proposal in a Comprehensive Reform

Leonardo Costa[1], Arlindo Cunha[2], Américo Mendes[1] and Miguel Sottomayor[1]
Faculty of Economics and Management[1], The Portuguese Catholic University, Portugal; Former Member of the European Parliament[2], Portugal

Current changes to the Common Agricultural Policy (CAP) started with the MacSharry reform of 1992 and continued with the Agenda 2000 reform of 1999. Several factors influenced these reforms: the Uruguay Round Agreement on Agriculture (URAA) and subsequent talks in the World Trade Organization (WTO), the budgetary costs of surpluses, concerns about the cost of CAP following eastern enlargement, a perceived need for rebalancing the CAP, and views about the negative impacts of agriculture on the environment. The CAP has not served well Europe's farmers or the agri-food industry. It provides unequal support among products, regions, and farmers, it imposes costs on the rest of society, it has exacerbated intensive agriculture's negative impacts on the environment, it does not help the EU's position in the WTO, and it created difficulties for EU enlargement. As a result it has been suggested that CAP should be replaced by CARPE, a Common Agricultural and Rural Policy for Europe, with the emphasis on the provision of environmental and rural services (Buckwell *et al.*, 1997).

Since the MacSharry reform, direct payments have been a central element of the CAP, providing compensation to farmers for the reductions in price support. Accompanying measures to support early retirement, to reforest agricultural land and to pay for the provision of agri-environmental services were also introduced. However, direct payments were not fully decoupled from production. Farmers must plant some crops or hold some animals to receive them. Under the WTO rules, the direct payments are classified as blue box, a category protected by the Peace Clause that will end by 2004. Thus, at the time of writing, the future of direct payments is uncertain, and the expectation is that they may be subject to greater restrictions. This expectation also results from EU eastern enlargement.

The cost of extending direct payments to the future Member States will be high. These problems add to pressure for further reform of the CAP[1].

Following on from the 1992 CAP reform, Agenda 2000 produced more of the same, except that CAP now has two pillars. The first pillar involves price support and direct payments for agricultural producers, and is responsible for approximately 84% of current CAP expenditure (EAGGF Guarantee and Guidance), and 90% of EAGGF-Guarantee expenditure. The remaining CAP expenditure (16% of the total, or 10% of EAGGF-Guarantee) goes to the second pillar, the most suited to support European agriculture's multifunctionality. The bond scheme proposal concerns CAP's first pillar and, in particular, its direct payments. It does not directly address CAP's second pillar support to farms and rural areas, although Swinbank and Tangermann, in Chapter 4, propose that it be 'embedded in a comprehensive reform of the CAP to better reflect the public's wider concerns about the environment and rural development'. Thus the bond scheme proposal can be seen as an alternative to the Transitional Adjustment Assistance (TAA) temporary element in Buckwell *et al.*'s (1997) CARPE, as it would ease farm adjustment in the transition from CAP to CARPE.

The bond scheme proposal

The bond scheme proposal advocates the removal of the *direct payments* of CAP's first pillar over a period of time (perhaps 15 to 20 years) and the transformation of the payment entitlements during this period into a system of tradable bonds. These bonds would help farms adjust in the transition from CAP to something such as CARPE. The bond scheme proposal implementation would imply:

1. Establishing a deadline for direct payments (that would finish after a set period of, say, 15 to 20 years).
2. Fixing the future level of annual payments in the transition period (15 to 20 years), removing the uncertainty of these payments in the period, and making payments eventually degressive in the period.
3. Decoupling payments from current cropland use, number of animals kept, and land (or farm) in the transition period and allocating entitlement to payments to individuals, who in turn would be free to sell (capitalise) entitlement to the future payments stream in the bonds market.

There are four main justifications for the bond scheme proposal. First, farmers are forced to plant some crops or hold some animals in order to receive the current direct payments, which reduce the economic value of the payments to them and to society. Direct payments interfere with farmers' decisions, and they are not helping farms adjust to changing market conditions or to a new CAP. Second, the future of direct payments is uncertain. It is uncertain because these payments may start to be subject to reduction requirements in the WTO. It is also uncertain because the eastward enlargement may undermine their sustainability. Currently direct payments are equivalent to 20% of EU farms' net value added. Third, direct payments do not contribute to correct CAP's unequal support across products, regions, and farmers: indeed their effect is quite to the contrary (Cunha, 2000, 2002). In the EU, 20% of farmers receive 80% of CAP support. Direct payments are linked to certain products only (cereals, milk, and meat) and to historical production, which is a burden for those regions and farmers less suited to produce these products or with low historical yields. Fourth, funds released could have more constructive uses in rural areas.

Multifunctionality

As pointed out by the European Commission, 'For centuries Europe's agriculture has performed many functions in the economy and the environment and has played many roles in society and in caring for the land' (Commission, 1997). They also claimed that 'The fundamental difference between the European model and that of our major competitors lies in the multifunctional nature of Europe's agriculture and the part it plays in the economy and the environment, in society and in preserving the landscape, whence the need to maintain farming throughout Europe and to safeguard farmers' income' (Commission, 1997). Although the term 'multifunctionality' generates considerable debate in international circles, particularly with regard to the mechanisms governments could, or should, deploy to secure the delivery of agriculture's multifunctional characteristics (see, for example, Mendes, 2002), most authors accept that agriculture has multifunctional characteristics. The OECD (1999, 2000) discussed multifunctionality; and Swinbank (2001) provided a more sceptical view. Table 5.1 provides a non-exhaustive list of non-commodity functions sustained by agriculture. The EU does not itself claim that 'food security' (listed in Table 5.1) is a multifunctional characteristic of agriculture. It may claim, however, food safety, animals' wellbeing, organic farming - none of which is listed in Table 5.1 - and products with local origin (listed in Table 5.1 as Cultural heritage) as multifunctional characteristics of agriculture.

The two key elements of multifunctionality recognised in OECD (2000) are '(1) the existence of multiple commodity and non-commodity outputs that are jointly produced by agriculture, and (2) the fact that some of the non- commodity outputs exhibit the characteristics of externalities or public goods, with the result that markets for these goods do not exist or function poorly.' Some argue that supporting agricultural production using production-linked payments may enhance its multifunctional character. In other words, production-linked payments may be necessary to obtain socially desired non-commodity outputs.

Figure 5.1 shows the contrast between the distribution of the CAP's 1995 benefits (the budget and trade effects expressed as a percentage of GDP, and then ranked) and GDP per capita among the fifteen Member States of the EU. The further Member States are from the diagonal, the more perverse is the positive or negative distributive effect. In particular, there were three Member States with below average income that were losers from the CAP - Portugal, the UK, and Italy - and two Member States with above average income that were gainers from the CAP: Denmark and France. This distribution has been little changed by the MacSharry reform (Buckwell *et al.*, 1997), or by Agenda 2000's CAP reform (Cunha, 2000, 2002).

Table 5.1. Non-commodity effects of agriculture

Environmental/Positive	Open space, Scenic vistas, Isolation from congestion, Watershed protection, Flood control, Groundwater recharge, Soil conservation, Biodiversity, Wildlife habitat
Environmental/Negative	Odour, Nutrient/pesticide runoff, Watershed protection, Flood control, Soil erosion, Biodiversity loss, Wildlife habitat
Food Security	Elimination of hunger, Assure availability of food supply
Rural Development	Rural income and employment, Viable rural communities
Social	Traditional country life, Small farm structure, Cultural heritage

Source: adapted from Bohman *et al.* (1999).

Whilst Fig. 5.1 illustrates the impact of the CAP on the economy as a whole, Fig. 5.2 focuses on its impact on farmers. For each country (and EU-15) the bar records the Agricultural Value Added per agricultural work unit (AWU) (i.e. the amount available to reward the labour, capital and land used in agriculture). It is made up of three components: an estimate of the degree of market support provided by the CAP (in black); a measure of the subsidies less taxes (in grey), which is dominated by direct payments under the CAP; and a residual component (in white) which might be termed the net value added at parity (i.e. world market) prices. In Ireland and Sweden, for the year recorded, NVApp was only just positive, whilst for Finland it was -€1,276. Figure 5.2 shows that some farmers clearly gained little from the CAP. Hence Cunha's (2002) earlier comment that Portuguese 'farmers receive on average five times less than their European fellows, in spite of having less than one third of their income'.

The unequal distribution of CAP benefits across products, regions, and farmers in the EU is not driven by multifunctionality. Instead, it is driven by the historical precedent of greater support to some sectors: cereals, beef, and milk.

The production-linked direct payments introduced by the MacSharry reforms, enhanced by Agenda 2000, based upon past policy, simply sustained the inequality of support. Together with quantitative restrictions, this has been the political choice so as to avoid market surpluses and subsidies to exports, the latter being largely restricted by WTO rules. As they are, direct payments do little for European agriculture's multifunctionality. The bond scheme proposal advocates the removal of these payments.

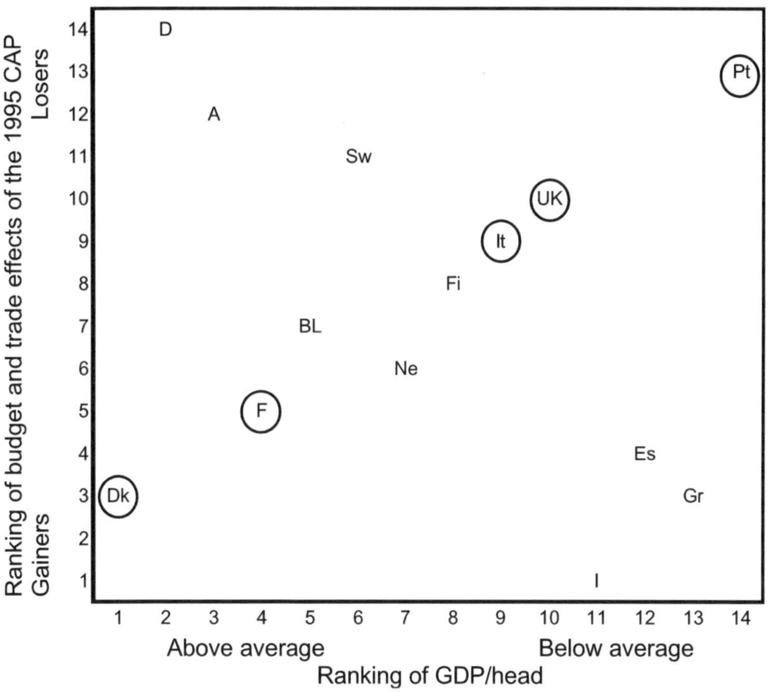

Fig. 5.1. The equity of the CAP. Distribution of the benefits of the CAP 1995. Source: Buckwell *et al.* (1997)

The URAA recognises countries' rights to pursue domestic agricultural policy objectives, including those under the rubric of 'multifunctionality' (Bohman *et al.*, 1999). WTO rules do not conflict with the sovereign right of countries to have objectives for their agricultural sectors. However, they require that policies aimed to achieve those objectives are minimally trade distorting (Bohman *et al.*, 1999). This is the over-arching condition of minimal trade distortion that applies in the green box. It will not be easy to expand the green box provisions in order to fully embrace the concept of multifunctionality as many countries want to tighten up on these provisions. It will be particularly difficult to devise new green box provisions that incorporate production-linked payments as a mechanism to secure the delivery of agriculture's multifunctional characteristics. Mendes (2002) reviews alternative mechanisms to secure the delivery of agriculture's multifunctional characteristics.

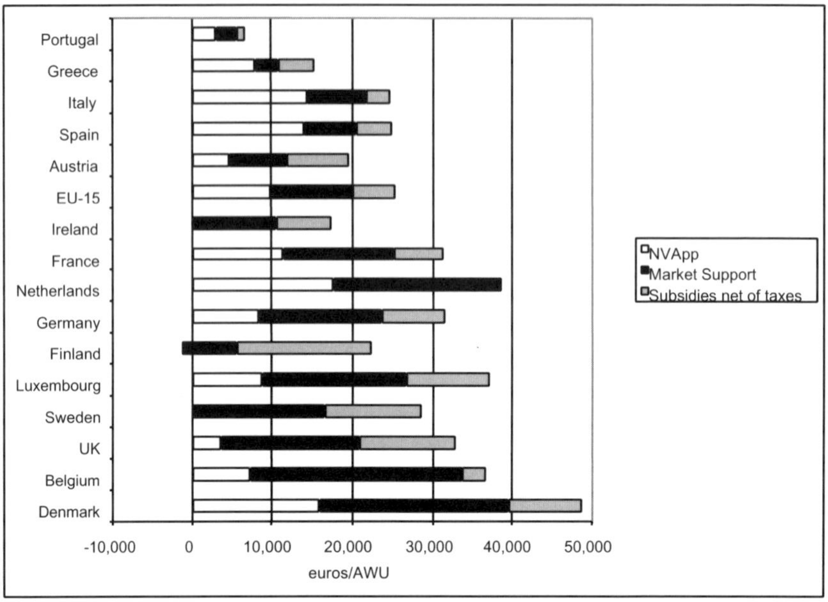

Fig. 5.2. Agricultural Value Added per AWU ranked by total support per AWU N.B. for Finland NVApp = -€1,276. Source: derived from data supplied by the Ministério da Agricultura, Desenvolvimento Rural e Pescas (MADRP) (based on Eurostat data), year uncertain

CARPE

In its seminal document entitled *The Future of Rural Society*, the European Commission described the European rural world as representing respectively 50% and 80% of the EU-12 population and territory (Commission, 1988). In a 1999 study from the European Parliament, these numbers were 31% and 76% of the EU-15 population and territory (Cunha, 2000). For the Mediterranean Member States of the Union, these percentages are, in general, higher.

The European rural world has a diverse number of economic activities: agriculture, handicrafts, small and medium businesses (manufacturing, trade, and services) (Commission, 1988). However, agriculture is not necessarily the main economic activity as, according to Buckwell *et al.* (1997), agriculture now contributes less than 2% of the EU-15's GDP and employs just over 5% of its workforce. Nevertheless, it still occupies large portions of territory, and it also supports the existence of many other activities. In Europe, rural landscapes have been agricultural landscapes for more than two thousand years.

The European Commission (Commission, 1988) has identified the existence of three types of problems in EU rural regions:

1. Regions subject to pressures of modern evolution, as with the commuting areas around large cities, or in coastal areas.
2. Regions that need to diversify their set of economic activities to stop rural decline, as with many regions in the Mediterranean Member States.
3. Marginal regions, which will need continuous support to keep a minimum population occupying the territory.

At a slow pace, the Union is trying to find answers to these problems. The most significant steps have been the so-called Delors Packages I (1989-1993) and II (1994-1999) to increase spending on the regional and social funds (Cunha, 2000). Pilot rural development programmes, such as the LEADER programme, which was administered by DG VI (Agriculture), and the current LEADER+ programme must also be noted. Agenda 2000 was portrayed as a serious attempt to co-ordinate, simplify, territorialise (or regionalise), and decentralise policy. For example, the second pillar of CAP is under a single regulation (Regulation 1257/99), and its implementation is decentralised by Member State (Council, 1999). However, the EU is still far from having a set of coherent policies for its rural and urban worlds. The EU is still exploring the functions it wants European agriculture to play in its different rural territories, as well as trying to establish the policies most suited to support them.

Buckwell *et al.* (1997) proposed that CAP should be replaced by CARPE (a Common Agricultural and Rural Policy for Europe). CARPE did not deal with many issues relevant to rural areas, such as health, housing, education, and so on. Instead, Buckwell *et al.* (1997) argued that 'The special role for a CARPE is to contribute to those aspects of rural development which relate directly or indirectly to land use'. And again, that 'The objective of a common agricultural and rural policy for Europe is to ensure an economically efficient and environmentally sustainable agriculture and to stimulate the integrated development of the Union's rural areas'. Three enduring elements characterised the CARPE proposal: market stabilisation (MS), environmental and cultural landscape payments (ECLP), and rural development incentives (RDI). A fourth, temporary, element was transitional adjustment assistance (TAA).

The first (enduring) element of CARPE, MS (market stabilisation), used CAP's commodity regimes to provide a residual safety-net of price support for commodities subject to wide market fluctuations. All other market supports would be removed, including direct payments. MS would be the only enduring element of CARPE that remained with a sectoral or commodity focus. The other two enduring elements, ECLP (environmental and cultural landscape payments) and RDI (rural development incentives), would be territorially defined and regionally administered. Together, the ECLPs and RDIs could be viewed as extended versions of CAP's second pillar, as inaugurated by Agenda 2000.

The second (enduring) element of CARPE, ECLP, contains payments to farmers for the provision of environmental and cultural landscape services. The principle of this element of CARPE was that farmers provide goods and services for which the market does not reward them. Theoretically, these payments would aim to avoid an under-provision of environmental and cultural landscape services. According to the Polluter Pays Principle (PPP), farmers could be obliged to avoid environmental damage and to comply with pre-established environmental standards, without receiving any payment for income losses. This is what the authors called Tier 0.

Thus, farmers should be rewarded for environmental services only when environmental objectives conflict with their established rights in resource use (land, water, etc.). Buckwell *et al.* (1997) defined two tiers for these rewards. Tier 1 would target and reward individual farming systems with a high nature value, i.e. farming systems which, when appropriately operated, have the potential or capability of delivering nature services over wide tracts of land. Tier 2 would target directly the territorial provision of environmental and cultural landscape services, instead of individual farms. Thus Tier 2 ECLPs would embrace the protection of high-interest eco-systems, wetlands, habitats for specific birds or other fauna, and the preservation of valued physical features in the rural landscape[2].

The ECLP element of CARPE could be viewed as an extended version of Agenda 2000's current agri-environmental measures and Less Favoured Areas (LFA) schemes. According to Agenda 2000, adoption of Best Management Practices (BMPs) is obligatory in areas vulnerable to nitrate pollution and is not the object of compensation. In non-vulnerable areas, the LFA scheme rewards farmers for adoption of BMPs (or low input farming systems), while the agri-environmental measures pay farmers for environmental services that go beyond BMPs.

The third (enduring) element of CARPE, rural development incentives or RDIs, concerns all aspects of rural development, including agricultural development. The emphasis is on stimulating opportunities for non-agricultural uses for farm resources and opportunities for resources released from agriculture. It corresponds to the existing structural measures of CAP's second pillar, and to pilot programmes such as LEADER and LEADER+.

The fourth (temporary) element of CARPE is the TAA (transitional adjustment assistance). Transition from CAP to CARPE imposes costs on existing beneficiaries under the CAP. Buckwell *et al.* (1997) proposed that the direct payments introduced in 1992 would be converted into transitional adjustment assistance and gradually reduced as support switched increasingly to environmental payments and rural development. 'Much of the transition assistance is to provide a cushion whilst farmers receive and digest the message that society will pay market prices for the marketable products, and reasonable prices for the non-market services they provide but not more than this.'

According to Buckwell *et al.* (1997) 'TAA should be decoupled from production. It should be non-distorting as regarding competition, and recipients should respect environmental conditions'. This latter aspect is reflected in the cross compliance provisions of Agenda 2000. Buckwell *et al.* (1997) suggested that TAA payments could be per hectare, per farmer or per annual work unit but should in no way relate to current decisions about production, resource use, current prices, or even be dependent on whether the recipient is still farming. However, having these characteristics, it is not clear how the authors could insist that recipients should respect environmental conditions. That is, how can cross compliance be a requirement for farmers receiving the TAA? The TAA would smooth the transition from CAP to CARPE by compensating farmers. Figure 5.3 illustrates CAP budget shares in 1990 and 1996 and suggested budget shares for CARPE in 2002 and 2008, as proposed by Buckwell *et al.* (1997).

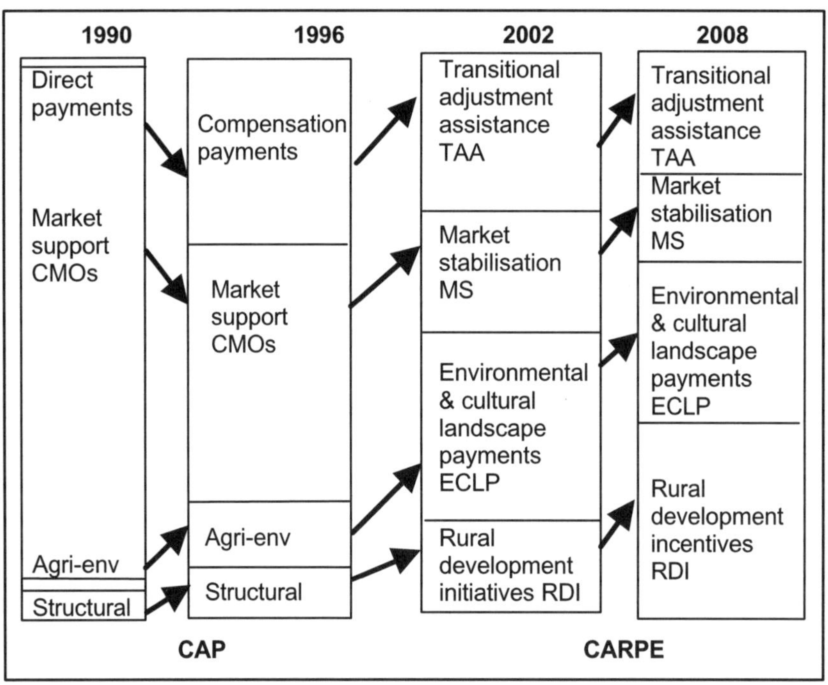

Fig. 5.3. Budget shares of CAP and CARPE elements. Vertical scale represents notional budget shares. Redrawn from Buckwell *et al.* (1997).

Many of Buckwell *et al.*'s (1997) CARPE proposals have been considered in Agenda 2000. However, the CAP's second pillar measures are still bolted-on extras to the core measures, which concern price and income support (CAP's first pillar). Buckwell *et al.* (1997) asserted: 'This balance has to be reversed'. The Agenda 2000 CAP reform has not reversed it.

Swinbank and Tangermann, in Chapter 4, have proposed that CAP's first pillar direct payments should be replaced by a system of tradable bonds, issued to farmers according to the current allocation of these payments, fully decoupled from production, and with a limited time horizon. The bond scheme can be seen as a proposal for the temporary element of CARPE (TAA). The proposal does not concern the enduring elements of CARPE (MS, ECLP, and RDI), although funds released could be allocated to support those enduring elements, reversing the balance of CAP. The bond scheme can be seen as a proposal to ease farm adjustment in the transition from CAP to something like CARPE.

By establishing a time limit for direct payments and by fixing their future level in advance, the proposal removes farmers' current uncertainty on the future of these payments. It does not remove, however, farmers' uncertainty about the future of CAP's, or CARPE's, enduring elements. Agenda 2000's second pillar showed, however, the type of support farmers might expect in the future from CAP. The bond scheme proposal would allow the smooth release of funds from CAP's first pillar for transfer to second pillar payments. Also, it would avoid farmers from the eastern European countries, on the verge of entering the Union, from receiving policy support for the wrong reasons[3], and it would make stronger the EU's negotiating position in the WTO.

Prospects for the future

Two future CAP reforms are in train: the mid-term review of Agenda 2000 launched in July 2002, approved in 2003; and the likely more substantial reform to take place in 2006 when the existing Berlin arrangements and the financial perspectives expire.

Whatever the content and substance of these reforms might be, there is a broad consensus on the need to reinforce rural development and give the CAP better balance between its policy goals. Commissioner Franz Fischler has recognised this need on many occasions, for example at the informal meeting of the Council of Agriculture Ministers in Murcia: 'we need to make an effort to reinforce rural development and allocate more resources to support a more environment friendly agriculture, to promote food quality and, at the end, to get a more competitive agriculture'[4] (Fischler, 2002). The European Commission's proposals for the mid-term review are fully consistent with Commissioner Franz Fischler's previous declarations.

Although presented by Swinbank and Tangermann, in Chapter 4, as a simple, incremental change, the bond scheme proposal could be interpreted as a *radical reform*. Radical reforms are not, however, a likely outcome of the EU's political decision making process, due to the power relationships among Member States and the redistributive effects they imply. Any policy reform tends to be incremental (Kay, 1998), especially when it may involve more budget resources (which is not the case of the bond scheme by itself), as will be the case with eastern enlargement. Embedded in a more comprehensive reform, the bond scheme can be seen as an appropriate instrument that would ease a substantial move in the CAP for the following reasons:

1. It is a policy instrument typically suited to smooth the transition from CAP to something like CARPE.
2. It is a policy instrument capable of implementation under many models of reform (more or less radical) aimed, however, at achieving the broad objectives already outlined.
3. It ensures transitional adjustment assistance to farmers, in the transition from CAP to an alternative form of agricultural and rural policy such as CARPE.
4. It allows farmers to release resources for use in more competitive agricultural production.
5. It smoothly releases funds from CAP's first pillar that could be better used in new policies in support of European agriculture's multifunctionality (CAP's second pillar), whilst leaving intact safety-net support for market stabilisation (CAP's first pillar).
6. It helps to correct CAP's historical unequal support among products, regions, and farmers.
7. It avoids farmers from the eastern countries entering the Union receiving support for the wrong reasons.
8. It makes the EU's negotiating position in the WTO stronger.

Conclusions

With Agenda 2000 the CAP gained two pillars. The first pillar contains all market and income support policies, including the MacSharry reform's direct payments. The second pillar contains environmental and rural development policies, including the LFA scheme and the agri-environmental payments. The first pillar provides support to producers of agricultural products. The second pillar aims to support other services provided by agriculture to society, namely multifunctionality.

CAP's first pillar is at stake in the WTO. It constitutes the bulk of CAP expenditure (90% of EAGGF-Guarantee expenditure) and its policies are in the contentious 'amber' and 'blue' boxes. Within the EU, the first pillar is at the heart of CAP's unequal distribution of benefits among products, regions, and farmers; it does not support multifunctionality; and it is not helping farms adjust to a new CAP or CARPE.

The bond scheme proposal concerns the direct payments of CAP's first pillar. It does not concern CAP's market price stabilisation policies, CAP's second pillar, or CARPE's enduring elements. The bond scheme proposal can be seen as the conversion of direct payments into temporary adjustment assistance to ease farm adjustment in the transition from CAP to CARPE. By establishing a time horizon to direct payments and by fixing them in advance, the bond scheme proposal removes uncertainty on the future of these payments. However, uncertainty persists on the future of CAP's, or CARPE's, enduring elements. Agenda 2000's second pillar shows, however, the type of support farmers might expect in the future. The bond scheme would allow a smooth release of funds from CAP's first pillar to second pillar payments.

Notes

[1] According to Cunha (2002), 'the Doha declaration is loose enough to allow for the continuation of the status quo for a few years more, with the real negotiations not starting before the beginning of 2004; while the likely EU decision to phase in the existing direct aids to the future Member States of Central and Eastern Europe (CEECs) in a carefully staged manner removes the budget pressures up to 2006'.

[2] In Tier 1, targeted farms may coexist with non-targeted farms in the same territorial unit while in Tier 2 all farms of the territory to be protected are targeted. That is, Tier 2 territories seem to be protected areas.

[3] That is, replicating in the Member States entering the Union the historical precedent of greater support to some sectors: cereals, beef and milk. In other words, giving these Member States mainly current first pillar support when all Member States should be receiving mainly current second pillar support.

[4] Own translation from the French text.

References

Bohman, M.J., Cooper, J., Mullarkey, D., Normile, M.A., Skully, D., Vogel, S. and Young, E. (1999) *The Use and Abuse of Multifunctionality*. Economic Research Service, USDA, Washington, DC.

Buckwell, A., Blom, J., Commins, P., Hervieu, B., Hofreither, M., von Meyer, H., Rabinowicz, E., Sotte, F. and Simpsi Vîna, J.M. (1997) Towards a common agricultural and rural policy for Europe. *European Economy* Reports and Studies No 5. CEC, Luxembourg.

Commission of the European Communities (1988) *The Future of Rural Society*. COM(88)501. CEC, Brussels.

Commission of the European Communities (1997) *Agenda 2000. The future for European Agriculture*. Directorate-General of Agriculture (DG VI), CEC, Brussels.

Council of the European Communities (1999) Council Regulation (EC) No 1257/1999 of 17 May 1999 on support for rural development from the European Agricultural Guidance and Guarantee Fund (EAGGF) and amending and repealing certain regulations. *Official Journal of the European Communities*, L160.

Cunha, A. (2000) *A Politica Agricola Comum e o Futuro do Mundo Rural*. Plátano Edições Técnicas.

Cunha, A. (2002) What can we expect from the CAP Mid Term Review? *Agra Europe*, 25 January 2002, A/1-A/2.

Fischler, F. (2002) Entretien avec le Commissaire Européen de l'Agriculture, le Developpement Rural et la Pêche, au terme de la reunion informelle des Ministres de l'Agriculture, à Murcie. Bulletin España 2002. http//.ue2002.es/actualidad

Kay, A. (1998) *The Reform of the Common Agricultural Policy. The Case of the MacSharry Reforms*. CAB International, Wallingford, UK.

Mendes, A. (2002) *The Common Agricultural Policy's Second Pillar: the Role of Public Support for Voluntary Collective Action*. Working Paper No 3 of the Bond Scheme Project. Centre for Agricultural Strategy, The University of Reading, Reading, UK.

Organisation for Economic Co-operation and Development (1999) *Multifunctionnalite: Rapport sur la Situation et Proposition de Travail Future*. AGR/CA(99)2. OECD, Paris.

Organisation for Economic Co-operation and Development (2000) *Multifunctionality: Towards an Analytical Framework*. COM/AGR/APM/TD/WP(2000)3/FINAL. OECD, Paris.

Swinbank, A. (2001) Multifunctionality: a European Euphemism for Protection? FWAG Conference *Multifunctional Agriculture - a European Model*, 29 November 2001, National Agricultural Centre, Stoneleigh.

Chapter Six:

Why a Bond Scheme was not Adopted in 1992

Carsten Daugbjerg
Department of Political Science, University of Aarhus, Denmark

Introduction

Reforms of the CAP of the European Union have all been aimed at upholding the interventionist policy paradigm which was laid down when the policy was established in the late 1950s. Up until now, the most comprehensive reform was the introduction of direct payments in 1992 to compensate farmers for reductions in guaranteed minimum prices. Originally, the CAP was a high price policy system in which farmers were supported through market regulation mechanisms which maintained prices at a higher level than would an unregulated market. The MacSharry reform of 1992 changed the CAP to a policy which contained substantial elements of the low price model. The Agenda 2000 reform of 1999 moved the CAP even further in that direction.

EU agricultural policy makers have rarely discussed radical reform proposals which would reduce agricultural support substantially. There is, however, one short discussion of radical reform which deserves closer attention. During the discussions on the MacSharry reforms in 1991 and 1992, the then Danish Minister of Agriculture, Laurits Tørnæs, put forward a bond scheme as an alternative to the Commission's proposal on using direct headage and area payments to farmers as compensation for price cuts (Landbrugsministeriet, 1991). Under the bond scheme, the EU would issue bonds which would guarantee predetermined payments for a certain period. The bonds could be sold on the capital market if the farmer wanted a once and for all payment in order to adjust his production to a situation in which the level of subsidisation would be considerably lower. Alternatively, the farmer could keep the bonds and receive payments for a fixed period. After that period, direct payments would end.

The bond scheme would enable the EU to cope with some of the problems which the CAP faced in the early 1990s. Agricultural support would be decoupled from production and thus be considered a non-distorting support measure in the

GATT negotiations. This could revitalise the talks which had been suspended in December 1990 after failure to reach agreement between the US and the EC on agricultural trade. Decoupling also meant that agricultural spending increases could be controlled. The Commission estimated that the agricultural spending in 1991 would increase by approximately eight billion ecus, 'the highest annual increase ever recorded' (Manegold, 1991). A further 20% increase was expected in 1992 (Agra Europe, 1991d) with German reunification only accounting for about one-sixth of the increase (Agra Europe, 1991c). Unless the Community severely limited increases in spending, the ceiling of the Agricultural Guideline which the Heads of Government imposed in 1988 to control agricultural spending would be breached (Agra Europe, 1991a), thus causing a new budgetary crisis. Despite these merits of the bond scheme, the proposal was paid very little political attention. This chapter examines why.

The chapter[1] first describes the 1991 Danish bond scheme proposal. Then, it identifies a number of factors which explain the failure of the proposal. The factors selected for analysis are: the timing of the proposal; the coalition building efforts; the scope and extent of reform; bureaucratic interests; and the role of the French-German axis in EC/EU agricultural policy making. In the conclusion, it is assessed whether, and under which conditions, the bond scheme will be a politically viable option in future CAP reform.

The Danish bond scheme proposal of November 1991

Originally, the bond scheme was presented in the Land Use and Food Policy Inter-Group's (LUFPIG) report of November 1990. This group consisted of members of the key parliamentary committees of the European Parliament and the six largest political groups. The Inter-Group had asked six eminent agricultural and environmental economists to produce a report on the future of the CAP which was then commercially published (Marsh *et al.*, 1990)[2].

Although the idea of using bonds to compensate farmers was not really in line with the thoughts of the Agriculture Commissioner, it did play a minor role. In the Commission's final reform proposal to the Council, it was suggested that bonds could be issued to dairy farmers as compensation for cuts in milk quotas (Commission, 1991b). However, the Council of Agriculture Ministers refrained from enacting reform measures in the dairy sector; only a small cut in butter prices was accepted. Since the dairy sector was only subject to minor adjustments, the bond scheme did not really become an issue for discussion in the negotiations on the milk market regime. Nevertheless, at a meeting of the Special Committee on Agriculture (SCA), both Britain and Denmark supported the proposal. The main focus of the reform proposal was the cereals, protein and oil crops market regimes. The centre piece of the reform proposal was price cuts compensated by direct payments. These would be permanent and provide full compensation; however, with only partial compensation on larger farms (modulation).

In the Council session on 18-19 November 1991, the then Danish Minister of Agriculture, Laurits Tørnæs, suggested the bond scheme as an alternative to the Commission's proposal. In the note which outlined the general features of the proposal, it was argued that, in the long term, the Commission's proposal would not solve the income problems of farmers because direct payments would be capitalised into land prices and maintain the price level because compensation was to be paid not only to farmers in business but also future purchasers of farms. The purpose of the Danish proposal was to adjust the CAP to market conditions in order to maintain a competitive agricultural sector. The compensation payments were to equal the capital losses per hectare caused by price reductions. Payments could be provided by issuing bonds to farmers which entitled their holders to predetermined and guaranteed annual payments for a given period. Bonds were only to be issued to farmers currently in business, not farmers entering the industry. These bonds were to be tradable in the capital market. Thus, farmers were free to keep them and receive annual payments or sell them in the private capital market to receive a once and for all capital sum (Landbrugsministeriet, 1991).

Since the payments were decoupled from production, they would, according to the Danish government, definitely be a 'green box' measure in the GATT negotiations on agricultural support (Landbrugsministeriet, 1991).

The fate of the bond scheme in the MacSharry reform

The MacSharry reform process will now be analysed in order to establish which political, procedural and institutional factors inhibited the bond scheme from being adopted in 1992. Theory is used as an analytical device to pinpoint the most important variables and relationships subject to empirical analysis.

Timing

Baumgartner and Jones (1993) argue that the way in which an issue is defined determines the contents of policy. The importance of issue definition in policy making was, in particular, recognised by Schattschneider (1975) who stated: 'the definition of the alternatives is the supreme instrument of power'. Thus, issue definition is not a technical process, but a highly political one which is dependent on the right political circumstances. Timing is an important factor if one wants to define or redefine an issue. It is difficult to redefine an issue if a competing definition has been accepted as the foundation of policy deliberations. The early stages of a policy process are usually the best time to influence issue definition. In EU policy making this is especially important (Peterson, 1995). As Peters

(1994) points out: 'The interests that can control this early definition are likely to be successful at the end of the process'.

Soon after the adoption of the budget stabilisers in 1988, it became clear to many EC agricultural policy makers that this new measure was insufficient to curb CAP expenditures. The Danish Minster of Agriculture realised this (interview: Tørnæs, August 2001), and so did the Commission (Kay, 1998). In fact, it was soon realised that the stabilisers, which lowered prices if certain production thresholds were exceeded, encouraged farmers to increase production to maintain their income, thus worsening the problem of overproduction (interview: Versteijlen, February 2002). Commission President Delors argued in early 1989 that a new reform of the CAP was needed and Agriculture Commissioner MacSharry began to prepare for reform soon after he took office in January 1989. The preparation was undertaken in a special and informal reform unit which involved only few people and was totally closed, even to the people inside DG VI (Kay, 1998; Daugbjerg, 1998; interview Meester, February 2002). Already in early 1990, Commission President Delors 'set out the initial lines of thought' on CAP reform (Ross, 1995). In October 1990, the Council discussed Agriculture Commissioner MacSharry's GATT offer which suggested a 30% cut in farm subsidies based on the 1986 level, and at that time it became clear that reform was on the agenda, and the direction of the reform content was starting to crystallise. The GATT offer triggered substantial opposition within the Council of Agriculture Ministers, and the Commission was forced to specify how farm incomes could be protected once a GATT agreement was reached (Daugbjerg, 1998). The Commission's next move was to propose direct compensation for price cuts (Swinbank and Tanner, 1996). The first Commission paper outlining this reform model was leaked to the press in December 1990 (Agra Europe, 1991b), and in July 1991 MacSharry presented the final version of the Commission proposal which laid down the new policy model in detail to the Council of Agriculture Ministers. He suggested that cereals prices were to be reduced to the expected world market level, which meant a 35% cut in guaranteed minimum prices. To compensate farmers for income losses and to cut down production, the Commission suggested a direct compensation scheme combined with set-aside requirements (Commission, 1991b).

The bond scheme was presented in November the same year. At that time the Commission proposal had been discussed for more than one year and the preparation of reform within the Commission had been going on for almost three years. This was particularly problematic for the Danish government because its proposal was fundamentally different from the Commission's proposal. While the bond scheme was based on the view that agricultural support should be phased out over time, the Commission's proposal attempted to uphold the support level by changing the way of supporting farmers, i.e. shift from price support to direct support. Therefore, the bond scheme would require a discussion on the basic principles of reform, and it was far too late to enter such a discussion (interviews: Bukman and Meester, February 2002). After a Commission proposal has been presented, it is *the* foundation of the discussions and the President of the Council aims to reach consensus among the Member States on the basis of it (interviews: Bukman, February 2002; Cunha, April 2002; see also Moyer and Josling, 1990). This is the traditional way of conducting negotiations within the Council of Agriculture Ministers. Ministers know that, and therefore they try to shape the details of the proposal to the benefits of their own farmers (interview: Cunha, April 2002). As a former Dutch Minister of Agriculture, Piet Bukman, pointed out: in this phase, 'procedures are playing a more important role than substance' (interview: Bukman, February 2002). If a minister attempts to block the Commission proposal and switch the discussions to alternative proposals, s/he might become marginalised in the reform process and lose influence.

Thus, when the bond scheme was presented, the Commission's reform proposal had already gained a political momentum and the minds of agricultural policy makers were already preoccupied with it. They did not devote much attention to alternative proposals because it was considered unrealistic that they would become the subject of reform discussions. Furthermore, there was a risk that pursuing alternative reform agendas, which were unlikely to attract the support of the majority, would result in isolation of the Member State pursuing such a strategy and hurt its interests. Undoubtedly, the Danes knew the rules of the game. They knew that it was important to stay in on the discussions on MacSharry's proposal which seemed to be the most realistic reform outcome in late 1991 and early 1992, given the decision making rules. Therefore, they devoted many resources to the Commission's proposal, and as a result the Danish attempt to win support for the bond proposal was not particularly forceful (personal communication: Mortensen, December 2001; interview: Meester, February 2002).

To sum up, the timing of the presentation of the bond proposal was not optimal. It was presented relatively late in the reform process and therefore it was not seriously considered. The main issue of the reform was already defined. To the Commission and the majority of Member States, the reform was a question of maintaining agricultural support at the existing level (interviews: Ottosen, September 2001; Cunha, April 2002), not a question of phasing out agricultural support.

Coalition building

Institutions matter in politics. As Schattschneider (1975) once stated: 'organization is the mobilization of bias: Some issues are organized into politics while others are organized out'. The way the EU decision making is organised has important implications for agenda setting. To switch reform discussions from the Commission's proposal to another alternative would require a substantial effort. The Treaty of Rome states that the Commission proposes and the Council of Ministers decides. This control over the agenda is a major power resource of the Commission. According to the Treaty of Rome, the Council can only adopt modifications of Commission proposals unanimously, which is almost impossible because usually the Commission has supporters within the Council. A qualified majority can only adopt alternative proposals from the Council of Ministers if the Commission endorses it. Therefore, the Commission takes an active part in Council negotiations and is determined to defend its power position as agenda controller (Hayes-Renshaw and Wallace, 1997; Swinbank, 1997).

Institutional rules are not the only factor empowering a commissioner. In agricultural policy making, the personality and political skills of the commissioner makes a difference (Grant, 1997). To influence policy, he (only men have been appointed Agriculture Commissioner) must 'establish ... credibility by showing he is in control of agricultural policy and can get things done. Since power is divided in the EC, this requires skill at political persuasion'. Further, he must be assertive (Moyer and Josling, 1990) and master the skill of political manoeuvring in a complex environment. These personal factors, combined with institutional rules of EU decision making, mean that the Agriculture Commissioner can forcefully pursue his policy objectives.

The bad timing of the Danish proposal meant that a considerable effort was needed to get the bond scheme onto the official Council agenda. MacSharry was generally regarded as having a stubborn nature and unlikely to change his proposals. As Grant (1997) points out: 'MacSharry's skills included a grasp of the complexities of his brief and a canny personality which mixed an undemonstrative nature with considerable determination and, when necessary, aggression'. The former British Minister of Agriculture John Gummer supports this characterisation of MacSharry by saying: 'Without MacSharry's drive and determination it must be doubted whether the (MacSharry) reforms would have succeeded, at least in the form in which they eventually emerged' (Gummer, quoted in Kay, 1998). Also the Danes felt the determination of MacSharry. He was unwilling to consider other alternatives and held on to his own proposal (interview: Ottosen, September 2001). He did not even recognise that the Danish proposal was an option after Tørnæs had attempted to revitalise it in February 1992. In the Council session of 2-3 March 1992 he said that the only reform proposal on the table was that of the Commission (Landbrugsministeriet, 1992)!

This determination, combined with the decision rules supporting the Commissioner's control over the agenda, meant that the Danish Minister of Agriculture had a very limited opportunity of winning sufficient support for the proposal to force MacSharry to propose the bond scheme.

There were, however, potential allies among the Member States. Belgium, Britain, Denmark and the Netherlands were against permanent direct support (Daugbjerg, 1998). This was probably the reason why Tørnæs felt that there was some support for the bond scheme within this group of Member States (interview: Tørnæs, August 2001). Nevertheless, only the British Minister of Agriculture, John Gummer, officially expressed limited support in a Council meeting in that he asked for further examination of the proposal. The Dutch Minister of Agriculture, Piet Bukman, who was President of the Council of Agriculture

Ministers in the second half of 1991 and thus chaired the Council meeting when Tørnæs presented the bond scheme proposal, did not use his position to put the proposal onto the Council agenda although he was sympathetic to the idea. The reason for this was that there was a general understanding among the majority of the Member States that the Commission proposal was the only basis for discussions. Therefore, a discussion on the bond scheme would not lead anywhere. The intention of the Dutch Presidency was to reach consensus on some parts of the reform, and a discussion on the bond scheme did not seem fruitful in that respect (interview Bukman, February 2002). However, in the meeting of the SCA the Dutch chair did try to put the bond scheme onto the agenda of the SCA, but there were no positive reactions to this, and therefore no further actions were taken (interview: Meester, February 2002).

Belgium, Britain, Denmark and the Netherlands formed a blocking minority. Surprisingly, the Danes did not lobby intensively within this group of Member States to build up support for the bond scheme. The proposal might have been discussed, but did certainly not dominate the discussions within the group. Rather, the four countries' positions on the MacSharry reform dominated the meetings (interviews: Bukman and Meester, February 2002; Ottosen, September 2001)[3].

Scope and extent of reform

Path dependent reforms are much easier to adopt than path breaking reforms. Path breaking reforms involve a shift in policy paradigm; that is, the overall goals of the policy are altered. Path dependent reforms can be undertaken within the existing policy paradigm (e.g. Hall, 1993). Path breaking reforms bring about uncertainty, which the established policy network formed around the policy wants to avoid. First, this uncertainty arises because it is difficult to know in advance what consequences a radically reformed policy may have. In other words, there is a risk of bringing about unintended policy consequences. For instance, the deregulation of agricultural policy may lead to a higher number of bankruptcies than foreseen. Second, it is unlikely that the network formed around an existing policy survives fundamental policy reforms. As a consequence, members holding central and powerful positions within the network may lose power when it is redesigned (Daugbjerg, 1999).

As mentioned above, the Danish bond proposal was an alternative to the Commission's proposal on compensating for price cuts with the introduction of permanent direct payments. The Danish proposal only referred to three arable market regimes and also operated with a system of price support to provide floor prices. Thus, its scope was somewhat narrower than the Commission's proposal which originally included the cereals, protein crops, oilseed, sheepmeat, milk products and beef market regimes. As to reform extent, the two proposals were different. While the Commission suggested permanent compensatory payments for price cuts, the Danish government wanted to phase them out over a certain, but not-specified, period.

During the debate on the bond scheme there seemed to be some confusion on the scope of the scheme. The bond scheme was first presented in the Land Use and Food Policy Inter-group's (LUFPIG) report of November 1990. Basically, the report advocated that support prices should be 'allowed to fall closer to world market levels' and the EC should 'issue farmers with bonds on which the Community would make annual payments for a certain number of years. The total amount of bonds would be based on the income loss expected to result from the cut in support prices' (Marsh *et al.*, 1990).

The LUFPIG proposal was more radical than the Danish proposal. It was broader in scope in that it included all major market regimes of the CAP. Similar to the Commission's and the Danish proposal, the LUFPIG proposal would retain the price support system as a safety-net, not as a measure to support farm incomes. The two versions of the bond proposal seem to have been discussed simultaneously, and this may have caused some confusion. For instance, in COPA, the EC farmers' association, the Danish proposal was presented in the Presidium and the Committee of General Experts, but the Working Party for Economics and Statistics analysed the LUFPIG report's version of the bond scheme (COPA, 1992; EF-udvalget, 1992a, b).

Although the Danish proposal was less radical than the LUFPIG proposal, it was still thought too great a step to take and it would be unusual in international negotiations which are usually characterised by small steps (interview: Ottosen, September 2001). The idea of phasing out substantial parts of agricultural support was seen as a major political problem by Agriculture Commissioner MacSharry. As he pointed out in a letter to the President of the Danish Agricultural Council in March 1992:

> 'Although the bond concept has its merits ... I would see important limitations on applying this approach on a large scale. Firstly there is the question of political acceptability. The bond scheme would imply ending support to most Community farmers after a given period. The farming community is concerned about the permanence of the proposed compensatory payments despite their guaranteed character, and would have great difficulties in accepting a time limited compensation system.'

This seems also to have concerned most Member States. Without mentioning the LUFPIG report and the bond concept, a British government report, prepared in 1995, analysed the radical version of the bond scheme. It was argued that it had a major political disadvantage. As the report concluded: 'Most fundamental, (the bond) option pre-supposes a desire to have agriculture operating in a free market. So far most Member States have shown no desire to move from policies which have prevailed over many years or to address the problems of the CAP in a fundamental way' (MAFF, 1995).

The fact that the adoption of the bond scheme would be a more extensive reform than would MacSharry's proposal caused some uncertainty about the consequences of applying bonds. Another major problem of the Danish bond proposal was to establish the actual compensation payments. Farm unions were worried about the possibility of estimating the appropriate compensation payments for the price reductions in both the direct payment model suggested by the Commission and in the bond scheme (COPA, 1992). The Agriculture Commissioner was also concerned about this problem, but only in relation to the bond scheme. He faced exactly the same problem in establishing the level of direct payments, but the Commission proposal allowed for future adjustments. As MacSharry stated in the letter of March 1992 to the President of the Danish Agricultural Council:

> '(Under the bond scheme)... compensation would have to be fixed once and for all, and no revisions would be possible. Thus any misjudgement in assessing the compensation initially would be perpetuated during the life of the system, for upwards of 20 years perhaps.'

Commission interests

Many studies have shown that public bureaucracies have interests of their own (e.g. Allison, 1971) and, perhaps most importantly, civil servants are in a position to pursue these bureaucratic interests because, to a considerable extent, they control the provision of expertise and information to ministers and commissioners (e.g. Moyer and Josling, 1990).

In EC agricultural policy making, the Commission has always feared renationalisation of agricultural policy because it was believed to have serious consequences for European integration (Commission, 1985). As Grant (1995) points out: 'the CAP has been seen as the corner stone of the integration process' (see also Kjeldahl, 1994; Keeler, 1996). Wilkinson (1994) has even argued that in the 1970s and 1980s, 'Those who dared even to consider the option risked their professional respectability'. Kjeldahl (1994) highlights the risks associated with CAP reform: 'A dismantling, even if only partial, of the CAP might be viewed as a signal of beginning disintegration of the EU'.

According to the former Danish Minister of Agriculture, Laurits Tørnæs, the then President of the Commission, Delors, feared that the bond scheme would lead to renationalisation of the CAP - a development he strongly opposed. The Director General of the DG VI, Legras, was 'vehemently against the proposal' and also feared that it would lead to renationalisation of the CAP, which he considered a political setback (interview: Tørnæs, August 2001). Adopting a moderate reform of the CAP which introduced direct payments did not involve a wholesale renationalisation of the CAP, though it contained limited elements (Wilkinson, 1994).

However, the bond scheme is not in itself a renationalisation of the EU agricultural policy; it is a phasing out of direct payments. It must be said, however, that the bond scheme may lead to renationalisation if Member States introduce national agricultural policies to support farm incomes in the absence of EU subsidies, and this is highly problematic for the CAP because it would distort competition among national farmers. Furthermore, over time, the bond scheme might threaten the interests of the Commission because it might change the power balance towards national agricultural administrations, which the Commission would consider highly undesirable. They would lose prestige and influence. Thus, invoking the renationalisation 'ghost' served as a means to underline the seriousness of abolishing the then major genuine common policy of the EC. Indeed, the MacSharry proposal *could* be seen as a bureaucratic gain as it would make the CAP even more demanding administratively and thus require hiring of additional staff. And so it did. For instance, in Denmark, the EC Directorate which carries out the day-to-day administration of the CAP market regimes increased its staff by 60% from 1991 to 1995 (Finansministeriet, 1992, 1996).

The French-German axis

Historically, France and Germany have been the pivotal Member States in EU agricultural policy making, and in agricultural reform processes the policy positions of these two states have been central to the reform outcome. Webber (1999) has examined conflicts over the creation of the CAP in the 1960s and over the MacSharry reform and the GATT Uruguay Round in the early 1990s, and has come to the conclusion on France and Germany that:

> 'In both periods, on almost all major issues, they initially took opposed positions. When and as long as their conflicts were not mediated and they remained divided, the outcome was a crisis and a decision-making deadlock ... If they were united from the outset in opposing a given project, ... their stance ... dictated the position taken by Brussels. Equally, once, despite initial differences, they reached a *modus vivendi* on a given project, ... Brussels took over their joint position' (Webber, 1999).

This pattern repeated itself during the Agenda 2000 negotiations in 1999, which further strengthens Webber's conclusion.

Traditionally, both states have been status quo orientated in reform processes, aiming at preserving the high price model of the CAP. Following Webber's (1999) argument, a reform proposal must be able to attract the support of Germany *and* France to be realised. If one of these two Member States, or both, remains opposed to a proposal, it stands a very limited chance of being adopted. However, a proposal need not win the support of both Member States from the outset. There are some indications that Germany is the most influential of the two in reform processes 'since it can exert influence on France to shift its position on agricultural policy positions' (Grant, 1997). This strong position may be associated with the fact that Germany is the single most important contributor to the CAP budget.

It is beyond any doubt that the Danish bond proposal suffered greatly from the lack of support of either Germany or France. During the reform process, there were no indications that these two Member States would support the bond scheme proposal, and this made it an unrealistic option. Not even Tørnæs' attempt to make the bond scheme a voluntary national option of the MacSharry reform could obtain the support of these two Member States (interview: Tørnæs, August 2001).

Conclusion and perspectives

The analysis of the Danish Minister of Agriculture's unsuccessful attempt to have the bond scheme adopted as the reform model in the early 1990s shows that the refusal had more to do with policy context and timing rather than the reform idea itself. Already when the bond scheme was presented, many political factors worked against the realisation of the idea.

First, it was presented relatively late in the reform process and at that time the Commission had already submitted its proposal. In accordance with the traditions in Council decision making, the Commission proposal was the focus of the negotiations. Second, the late presentation of the proposal required that the Danish Minister build up strong pressure within the Council to force the Commission to change its proposal, but he only made a half-hearted attempt to form a coalition which would support him. Third, the policy distance between the CAP at that time and the bond scheme was too great, meaning that ministers of agriculture were reluctant to consider such a step. Besides, two versions of the bond scheme were discussed and this confused the discussion. Fourth, the introduction of the bond scheme potentially threatened Commission interests because it would entail that a lot of administrative tasks disappeared. Finally, neither France nor Germany supported the idea of using bonds.

Since the adoption of the MacSharry reform in the early 1990s, many of the factors which inhibited the bond scheme from being seriously considered have changed in a direction favourable to the bond scheme, or can be changed in future reform rounds.

First, this latest academic work on the bond scheme has clarified the reform idea. It has now been developed into a stepwise, and not a full-scale, reform of the CAP which requires no budgetary increases (as outlined in Chapter 4).

Second, the MacSharry and Agenda 2000 reforms and the small-farmers' scheme have reduced the policy distance between the current CAP and the bond scheme, and therefore the bond scheme will now be considered a less radical reform option than it was in 1992. The MacSharry reform involved a transformation of price support into direct acreage and headage payments, and the Agenda 2000 reform further reduced price support and increased direct payments.

The transformation of price support into direct payments is the first step in moving the CAP to the bond scheme. Since the levels of direct payment generally are considered acceptable within the farming community, the problem of finding the correct compensation levels has been eliminated. In Swinbank and Tangermann's proposal, the existing direct payments are transformed into annual payments to bond holders. Thus, the 1992 CAP reform has, presumably unintended, solved one of the major difficulties of moving from direct payments to bonds.

The second step in moving the CAP to the bond scheme is to decouple direct payments from production inputs and introduce a flat rate payment. The Agenda 2000 reform introduced such flat rate payments by aligning payments on land planted with payments on land set aside and by eliminating the difference between payments on land planted with oilseed and cereals. The adoption of the small farmers' scheme has introduced decoupled payments into the market regimes of the CAP. It introduced a flat rate aid for farmers receiving less than €1,250 in direct aid payments. Farmers choosing to enter the scheme make one single application for the trial period 2002-2005. It is estimated that up to a third of all EU farmers would be eligible for the scheme (Agra Focus, 2001a; European Commission, 2001). The scheme will provide experiences and more certainty about the consequences of applying decoupled support measures, and this means that it would be easier to expand the scheme to all farm sizes. Although the experiences with decoupled payments, currently, are very limited, the EU has expanded the use of decoupled payments to all farm sizes in its mid-term review of the CAP.

Third, although the bond scheme is not in itself a re-nationalisation of the CAP, the re-nationalisation ghost cannot be invoked as forcefully as in 1992, since the EU has developed other common policies, most notably the common monetary policy. The CAP is no longer *the* only genuinely integrated common policy of the Union. The EMU is likely to become the major EU common policy and the Common Foreign and Security Policy will attract a lot of attention in the years to come. Therefore, within the College of Commissioners, opposition to the bond scheme may now have diminished since the phasing out of the direct payments of the CAP market regimes would not lead to disintegration within the EU.

Fourth, Germany is moving towards fundamental reform. It is no longer as status quo oriented as earlier. During the 1992-reform process, the German Minister of Agriculture fought for the traditional German position in EC agricultural reforms, namely maintaining prices at a high level at all costs, and solving problems of overproduction by quantitative restrictions (Hendriks, 1994). In a speech in London in July 2001, the German minister responsible for agriculture, Renate Künast, indicated that Germany wanted to change support instruments. As she said: 'I envisage a transfer of the various animal premia into a grassland premium as an interim step. ... In the long run this could evolve into a uniform basic premium per farm' (Künast, 2001). As argued above, such decoupling of support moves the CAP closer to the bond scheme.

Thus, the policy context is now more favourable to the bond scheme; although, a favourable policy context does not in itself promote a reform idea. It must be forcefully initiated, and this requires good timing. That is, it must be put onto the agenda in the early phases of the reform process, and it must be supported by a dedicated coalition which, initially, at least commands a blocking minority. A group of Member States which potentially could form such a coalition consists of Britain, Denmark, Germany, The Netherlands and Sweden (see Agra Focus, 2001b).

However, the changes in policy context are not all in favour of the bond scheme. Bureaucratic interests behind a continuation of the current CAP have been strengthened. The implementation of the MacSharry reform required additional staff to undertake administration of the direct payments and supply control regulations. The jobs and career opportunities of these people depend on the current CAP. Therefore, the bond scheme is not in their interest. When bonds have been issued, the need for administrative resources would decline dramatically, which is a threat to EU and national agricultural bureaucracies.

Further, new priorities in the debate on the future CAP seem to work against the bond scheme. Cross-compliance has been introduced and there is a strong desire to expand its use. As the Commission (2002) suggested in its proposal for a mid-term review: 'The Commission … proposes introducing a single decoupled payment per farm. … (Under this scheme) payments will be conditional on compliance with statutory environmental, food safety, and animal health and welfare standards (Cross-compliance)'. Under the bond scheme, annual bond payments 'cannot … be made contingent upon any 'cross-compliance' requirements relating to land use, animal welfare or environmental protection. Bond payments would be unconditional'. This is because the objective of the payments is to compensate for the capitalisation of past agricultural support. Integrating cross-compliance with the bond scheme would imply that certain requirements are linked to the receipt of bond payments. Thus, farmers' opportunities to adjust to a situation in which the support level would be significantly lower, as happens when the bonds expire, would be seriously weakened. It would not be possible to leave the land and use the capital value to enter other and more profitable businesses. Thus, cross-compliance only enables farmers to adjust partially. In Chapter 4, Swinbank and Tangermann contend that if politicians want to further land use, animal welfare or environmental protection by the use of agricultural support measures, they should apply earmarked subsidies and not link them to general support schemes. Such payments can be funded out of the budget savings by gradually decreasing bond payments over time.

To conclude, the fact that the bond scheme attracted little support in the early 1990s need not mean that it will be an unattractive option in future reform. The agricultural policy context has mainly changed in a direction favourable to the bond scheme. However, the favourable policy context will not in itself guarantee that the bond scheme will be put onto the agenda and be seriously considered. Contexts do not act, but constrain and facilitate certain actions! Political actors make policies. Therefore, members of the EU agricultural policy establishment must re-launch the idea in the very early phases of the reform process, and they must be able to utilise the institutional rules and personal factors present in EU agricultural policy making. If they play the political game skilfully, the bond scheme may become a serious reform alternative in the future.

Notes

[1] This chapter is a revised version of my paper at the 10th European Congress of Agricultural Economists, 28-31 August 2002, Zaragoza, Spain. I am very grateful to the Aarhus University Research Foundation for funding the research and to the Danish Ministry of Food, Agriculture and Fisheries, the Danish Council of Agriculture and the Danish Farmers' Union for allowing me access to their archives. Thanks are also due to the people, listed below, who I interviewed or corresponded with for the provision of valuable information.

[2] Professors John Marsh, Bryn Green, Brendan Kearney, Louis Mahé, Stefan Tangermann and Secondo Tarditi.

[3] It should be noted that Denmark took over the Presidency of the Council of Ministers for the first six months of 1993, and there were press reports that Tørnæs would use his position as President of the Council of Agricultural Ministers to reopen the debate over bonds (Agra Europe, 1993). However, a change of government quickly removed Tørnæs from this position.

Interviews/personal communication

Piet Bukman, former Dutch Minister of Agriculture and President of the Council when the Danish bond scheme proposal was presented.

Arlindo Cunha, former MEP, former Portuguese Minister of Agriculture and President of the Council at the time of the enactment of the MacSharry reform.

Gerrit Meester, Policy Adviser, Former Deputy Director, Department of International Affairs, Dutch Ministry of Agriculture.

Jørgen R. Mortensen, Researcher, Department of Agricultural and Resource Economics, University of Arizona, former Head of the Economics and Statistics Department, the Danish Farmers' Union.

Poul Ottosen, Permanent Secretary, Danish Ministry of Food, Agriculture and Fisheries, and former Head of the EC and International Affairs Department of the Ministry of Agriculture.

Laurits Tørnæs, Regional Mayor, former Danish Minister of Agriculture.

Herman Versteijlen, Head of Unit, European Commission, former member of Agriculture Commissioner Ray MacSharry's cabinet.

References

Agra Europe (1991a) 11 January, P/1.

Agra Europe (1991b) 18 January, E/2-E/7.

Agra Europe (1991c) 25 January, P/1.

Agra Europe (1991d) 15 February, P/1.

Agra Europe (1993) 8 January, E/4.

Agra Focus (2001a) July.

Agra Focus (2001b) November.

Allison, G.T. (1971) *Essence of Decision: Explaining the Cuba Missile Crises.* Little, Brown, Boston, Massachusetts.

Baumgartner, F.R. and Jones, B.D. (1993) *Agendas and Instability in American Politics.* The University of Chicago Press, Chicago, Illinois.

Commission of the European Communities (1985) *Perspectives for the Common Agricultural Policy.* COM(85)333. CEC, Brussels.

Commission of the European Communities (1991a) *The Development and Future of the CAP. Reflections Paper of the Commission.* COM(91)100. CEC, Brussels.

Commission of the European Communities (1991b) *The Development and Future of the Common Agricultural Policy. Follow up to the Reflections Paper, COM(91)100 of 1 February 1991.* COM(91)258. CEC, Brussels.

Commission of the European Communities (2002) *Communication from the Commission to the Council and the European Parliament. Mid-term Review of the Common Agricultural Policy.* COM(2002)394. CEC, Brussels.

COPA (1992) *Bonds as a Means of Compensating Farmers for Reduced Prices etc.* Draft, 17 March 1992. ES(92) 7. COPA, Brussels.

Daugbjerg, C. (1998) *Policy Networks Under Pressure: Pollution Control, Policy Reform and the Power of Farmers.* Ashgate Publishing, Aldershot.

Daugbjerg, C. (1999) Reforming the CAP: policy networks and broader institutional structures. *Journal of Common Market Studies* 37, 407-428.

EF-udvalget (1992a) *Referat af Møde I COPA's Præsidium.* 10 January 1992, EF-udvalget 12/92.

EF-udvalget (1992b) *Referat af Mode I COPA's Generalekspertgruppe.* 30 and 31 January 1992, EF-udvalget 29/92.

European Commission (2001) *Decision to Cut Red Tape for Small Farmers.* Press Release IP/01/882, 20 June 2001. European Commission, Brussels.

Finansministeriet (1992) *Forslag til Finanslov for Finansåret 1993.* Finansministeriet, Copenhagen.

Finansministeriet (1996) *Forslag til Finanslov for Finansåret 1997.* Finansministeriet, Copenhagen.

Grant, W. (1997) *The Common Agricultural Policy*. Macmillan, Basingstoke, UK.
Hall, P. (1993) Policy Paradigms, Social Learning and the State. The Case of Economic Policymaking in Britain. *Comparative Politics* 25, 275-296.
Hayes-Renshaw, F. and Wallace, H. (1997) *The Council of Ministers*. Macmillan, Basingstoke, UK.
Hendriks, G. (1994) German agricultural policy objectives. In: Kjeldahl, R. and Tracy, M. (eds) *Renationalisation of the Common Agricultural Policy?* Valby/la Hutte, Institute of Agricultural Economics/Agricultural Policy Studies.
Kay, A. (1998) *The Reform of the Common Agricultural Policy. The Case of the MacSharry Reforms*. CAB International, Wallingford, UK.
Keeler, J.T.S. (1996) Agricultural Power in the European Community. Explaining the Fate of CAP and GATT Negotiations. *Comparative Politics* 28, 127-149.
Kjeldahl, R. (1994) Reforming the Reform? The CAP at a Watershed. In: Kjeldahl, R. and Tracy, M. (eds) *Renationalisation of the Common Agricultural Policy?* Valby/la Hutte, Institute of Agricultural Economics/Agricultural Policy Studies.
Künast, R. (2001) Germany's ideas about a new European Agricultural Policy. Speech given by the Federal Minister of Consumer Protection, Food and Agriculture on the occasion of the International Conference *Where Next for European Agriculture*, London. 17 July 2001, http://www. verbraucherministerium.de
Landbrugsministeriet (1991) *Den Fælles Landbrugspolitiks Udvikling og Fremtid – Den Danske Delegations Bidrag*, Brussels. 20 November 1991, 9545/91.
Landbrugsministeriet (1992) *Bidrag til Reformdrøftelserne*. Landbrugsministeriet 2, afdeling, 1kt, J.nr. 91-203-1, 4 March 1992.
MAFF (1995) *European Agriculture: the Case for Radical Reform. Working Papers*. Prepared by the Economics (International) Division for MAFF's CAP Review Group. MAFF, London.
Manegold, D. (1991) EC Agricultural Policy in 1990/91. Growing need for real policy reform. *Review of Marketing and Agricultural Economics* 59(2), 98-129.
Marsh, J.S., Green, B., Kearney, B., Mahé, L., Tangermann, S. and Tarditi, S. (1990) *A Future for Europe's Farmers and the Countryside*. Land Use and Food Policy Inter-Group, European Parliament, Brussels.
Moyer, H.W. and Josling, T.E. (1990) *Agricultural Policy Reform: Politics and Processes in the EC and the USA*. Harvester Wheatsheaf, London.
Peters, B.G. (1994) Agenda-setting in the European Community. *Journal of European Public Policy* 1, 9-26.
Peterson, J. (1995) Decision-making in the European Community. *Journal of European Public Policy* 1, 69-93.
Ross, G. (1995) *Jacques Delors and European Integration*. Polity Press, Oxford, UK.
Schattschneider, E.E. (1975) *The Semi-sovereign People: a Realist's View of Democracy in America*. The Dryden Press, Illinois.
Swinbank, A. (1997) The CAP decision-making process. In: Ritson, C. and Harvey, D. (eds) *The Common Agricultural Policy*. 2nd Edition. CAB International, Wallingford, UK.

Swinbank, A. and Tanner, C. (1996) *Farm Policy and Trade Conflict. The Uruguay Round and CAP Reform.* The University of Michigan Press, Ann Arbor, Massachusetts.

Webber, D. (1999) Franco-German bilateralism and agricultural policies in the European Union: the neglected level. *West European Politics* 22, 45-67.

Wilkinson, A. (1994) Renationalisation: an evolving debate. In: Kjeldahl, R. and Tracy, M. (eds) *Renationalisation of the Common Agricultural Policy.* Valby/la Hutte, Institute of Agricultural Economics/Agricultural Policy Studies.

Chapter Seven:

Implementing a Bond Scheme

Alan Swinbank[1], Jonathan Little[1], Thomas Knapp[2] and Miguel Sottomayor[3]

Centre for Agricultural Strategy[1], The University of Reading, UK; Institut für Agrarökonomie[2], Georg-August Universität Göttingen, Germany; Faculty of Economics and Management[3], The Portuguese Catholic University, Portugal

In this chapter we explain in more detail how a bond scheme might be implemented, addressing some of the practical concerns that have been expressed in our wide consultations with stakeholders and in earlier debates about the introduction of a bond scheme.

Determining payment levels, base entitlements and payment profiles

As Daugbjerg (in Chapter 6) has indicated, in 1991 the bond scheme was seen as a radical policy change that involved not just a different system, but also the determination of payment levels and the distribution of entitlements to payments. But the MacSharry reforms also necessitated the determination of payment levels and base entitlements, and these have been translated into the Single Payment Scheme of the Fischler Reforms. Hence, the current suggestion that existing payments be transformed into a bond scheme presents far fewer problems than those faced by the authors of the LUFPIG report in 1990 (Marsh *et al*., 1990). The Member States are now in possession of a substantial database showing the IACS claims over many years and for many claimants there would be continuity in the level of payments.

Basing payment entitlement on past IACS claims is not, however, problem free. A minor irritant is that Member States continue to unearth fraudulent claims within the IACS data base: thus, in the UK, in 2000, penalties were imposed on 2.4% of claimants under the Arable Area Payment Scheme for making claims which were 3% (or 2 hectares) more than allowed (House of Commons, 2002).

More problematic is the fact that some farmers might feel that they had suffered rough justice in that payment entitlement had been determined on the basis of past IACS claims, and not on IACS claims that might have been made. For example, animal disease (e.g. foot and mouth), inclement weather, human frailty, etc. might mean that farmers had not claimed their maximum IACS payment in the reference period. Thus, the best three years out of five, with further discretion for arbitrators, might be a more appropriate rule. One farmer we consulted, whilst broadly welcoming the proposed bond scheme, felt that his past decision to switch from cereals to the cultivation of borage (on which IACS payments are not made) should not be held against him. He thought that a fairer system of allocating bond payments would be the area on which he was entitled to make IACS claims, that is, land in an arable rotation on 31 December 1991, and subsequently registered as eligible for arable area payments. Any such scheme might also involve a scaling down of farmers' claims, so that the aggregate allocation for the country (or region) was no larger than the actual IACS payment in the base period. In effect, this is what is happening under the regionalised version of the Single Payment Scheme.

Debate over the introduction of a Single Payment Scheme, to replace area and headage payments, in the mid-term review, focussed attention on those businesses that had divested themselves of, or acquired, assets during the reference period used to determine entitlements, and before the start of the new payment system. Farmers who bought arable land in 2003, for example, might have thought they had bought entitlement to future area payments; but if the new system were to be based on average IACS claims over a 2000/02 reference period they might find they had no entitlement to payments. One might argue that the sellers were astute, and that the buyers were not guaranteed that IACS payments would continue into the future; in short that the principle of *caveat emptor* should apply. Politically, however, this could be problematic, and so an alternative allocation mechanism had to be found. The Commission's legal services will have been anxious to avoid a repeat of the SLOM debacle in which milk producers who had been participating in non-marketing of milk, or dairy herd conversion schemes, at the time milk quotas were allocated, and who were initially denied quota, were subsequently allocated quota after the intervention of the European Court (Rodgers, 1998).

It might be that, on political grounds, it is decided that the level of payments should be abated either to provide funds for other policy initiatives (e.g. Pillar 2 expenditure) or to produce what might be seen as a more equitable pattern of compensation payments. As originally formulated, the Swinbank and Tangermann proposal (see Chapter 4) involved all recipients continuing to receive their full IACS payments for a number of years. However, other formulations are possible: for example there might be a cap on payments, as initially suggested in the mid-term review (Commission, 2002; at €300,000 per claim), and/or a reduced payment entitlement might be introduced for recipients of larger IACS sums. For example, under the Single Payment Scheme, claims of less than €5,000 will, in effect, be exempt from 'modulation'.

However, once payment entitlement was determined we would suggest that all bonds then have the same payment profile. One format in which bonds could be issued is in €1 *nominal* payment entitlements. Recipients might have received bonds for, say, €1,050 or €300,000, depending on their previous IACS claims. *All* bond holders would then be entitled to the same payment profile: say 100% of the nominal rate for years one to five, then 80% in year six declining to 20% in year nine, the last year of payment. Any attempt to allocate different payment profiles to sub-categories of bonds would reduce their liquidity on the stock market because it would create a number of distinct bonds, each with a lower market capitalisation than would be the case with a single bond.

Farmers or landlords?

With the proposed bond scheme, entitlement to future *compensation* payments under the bond would most readily be based upon IACS claims in a previous period. Under most circumstances the claimant would thus be the farmer rather than the landowner. However, the decoupling of payments from the land, and the creation of a paper asset (the bond), will tend to depress land (and other farm asset) prices whilst embodying in the bond the net present value of the expected stream of annual payments. It is, of course, not just farm returns that impact upon farmland prices, and many of these other factors are location specific. Consequently it is difficult to predict, in any particular farm circumstance, what the impact of the bond scheme would be on land prices. Economists at the British Ministry of Agriculture, Fisheries and Food have however suggested that 'around half the value of agricultural land represents the capitalised value of support' (MAFF, 1995). Other authors have, however, suggested rather larger figures: Frandsen *et al.* (2002), in modelling the 'elimination of all domestic support in the EU', suggested that land prices across EU-15 would fall by 78%, with the UK at this average EU level (*op cit.*, Table 7). In Portugal land prices were predicted to fall by a massive 95% following the elimination of all support, which presumably implies that significant areas of land would go out of production. However, the 'elimination of all domestic support' would still involve an annual expenditure of €4.1 billion on export subsidies, and the collection of €1.1 billion of taxes on agricultural imports, casting doubt on the internal consistency of the analysis.

The ownership and control of farmland varies across the EU. Table 7.1 gives a first, very crude, impression of the divergence of EU experience. It indicates that the percentage of agricultural land farmed by the owner ranges from 32% in Belgium to 87% in Ireland. However, this only gives part of the picture. Ravenscroft *et al.* (1999) indicate that apart from Spain, with its 'absentee landlords with substantial holdings (*latifundismo*)', the UK is fairly unique in having a discrete landlord-tenant system, with arms-length business transactions between tenant and landlord. Elsewhere, ownership of land and other farm assets is often retained within the family, and 'for the majority of European nations, the letting of private land outside the family is of relatively minor importance'.

Furthermore, Ravenscroft *et al.* (1999) suggest that, faced with restrictive national laws constraining tenancies and the ownership of land, many other forms of share farming agreements and other joint ventures (including partnerships, contract farming, and share-flocking, for example of sheep in the Lake District) have emerged in most Member States.

Table 7.1. Percentage of agricultural area owner-farmed, 1997

	%
Belgium	31.9
Denmark	75.1
Germany	37.0
Greece	73.8
Spain	72.3
France	34.9
Ireland	86.7
Italy	78.1
Luxembourg	46.5
Netherlands	71.7
Austria	77.2
Portugal	69.6
Finland	80.2
Sweden	54.4
UK	65.2
EU-15	59.0

Source: European Commission (2002b).

The regulation of landlord-tenant relations is not an EU competence: the European Court has confirmed that 'the legal relations between landlord and tenant remain governed by the national laws of the member states' (Rodgers, 1998). As the contributions in Bartélemy and David (2001) make clear, there is considerable variation between Member States in the way milk quotas and the entitlement to receive suckler cow and sheep premia have been treated within this framework. Thus our conclusion is that it would be inappropriate for the EU to determine how a bond scheme should fit into landlord-tenant relations, or for us to arbitrate on how entitlement might be split between landlords and tenants. Some further pointers are, however, in order.

For example Rodgers (1998) has pointed out that (in the UK at least) milk quota is linked to the land, whereas livestock quotas are linked to the producer. This means that in the case of milk quota the landlord's interests tend to prevail, and tenants cannot sell milk quota from the farm without the landlord's permission, although the *Agriculture Act 1986* protects the tenant's interest and allows for the departing tenant to be compensated for any 'betterment and enhancement' of the farm reflected in its quota allocation (Rodgers, 1998). By contrast, it would appear that tenants are free to sell entitlement to suckler cow and sheep premia off the farm, or to remove the entitlements from the farm on the termination of the tenancy, which will 'prejudice the landowners interest, and may depress the rental value of the holding', unless the landlord has made 'specific provision for livestock quota in the tenancy agreement' (Rodgers, 1998).

Although a bond scheme was first proposed in 1990, and suggestions for decoupled compensation payments go back to the 1960s, in our consultations with stakeholders, we found no evidence to suggest that landlords have considered including provisions within tenancy agreements to deal with a policy change of this order.

Bonds, and their market capitalisation

If the proposed bond scheme were adopted, existing IACS claimants would receive an annual entitlement to receive a compensation payment for a specified number of years, and this entitlement could be sold or otherwise transferred. The annual payments would be unconditionally guaranteed by the EU, and so would be comparable to a conventional government (EU) bond. The annual payments would mimic the annual interest payments on a government bond, but there would be no repayment of principal on expiry of the bond as its nominal value would be zero. Thus the proposed scheme might be said to create *coupon strips* rather than *bonds*, akin to the financial assets created by *stripping*. Stripping has become a common practice in these markets. As the Deutsche Bundesbank (2002) remarks: 'Stripping, the separate trading of the registered interest and principal of securities, has been possible in the case of certain ten and 30-year Federal bonds since the middle of 1997. ... The reconstruction of the original bond, the underlying bond, from the coupon strip and the principal strip is also possible; however, only credit institutions, financial service institutions, securities trading firms and securities trading banks may do this with their own bond holdings'. Thus there is nothing inherently new in the bond scheme proposal.

The European Investment Bank (EIB), one of the EU's institutions, borrows on capital markets by issuing gilt-edged stock. For example, on 19 September 2002 the EIB issued a prospectus for the sale of €5 billion of EARNs (Euro Area Reference Notes), at an issue price of 99.965%, an annual interest payment of 3.50% on the nominal sum (paid annually on 15 October in arrears) and a maturity date of 15 October 2005 (EIB, 2002). As of 12 September 2002, the EIB had EARNs outstanding as recorded in Table 7.2. The Member States are the members of the EIB, and the ultimate guarantors of the gilt-edged stock on issue. Thus the bond scheme would simply create an additional financial asset, backed by the EU, traded on the euro bond market.

Based on the year 2000 payments, bonds would initially be issued to 3.1 million IACS claimants, and the initial annual payment would average €6,600. There would, however, be a very large number of small payments, and a relatively few large sums, as can be seen in Fig. 7.1: 37% of claimants received less than €1,250 and obtained 3% of the funds, whilst 0.02% of claimants received in excess of €500,000 and pre-empted 2% of the funds.

Although small payments cannot necessarily be equated with small business size (a very large farm business, specialising in fruit and vegetable production for example, may only have a small IACS claim), it would be our guess that the majority of farmers in receipt of relatively small sums would choose to hold their bond, rather than sell it on the capital markets. Their inexperience of dealing with financial instruments would, probably, prompt caution. However, even if a large number of recipients of small amounts were to hold their bonds, over 55% of the annual payments would lie between €10,000 and €100,000, and could be traded for a sizeable capital sum for reinvestment elsewhere (on the farm, in another business venture, or in stocks and shares).

Table 7.2. EARNs outstanding, 12 September 2002

Coupon	Redemption	Nominal value (€)
4.50%	15 February 2003	3,359,796,072
5.25%	15 April 2004	6,188,037,320
3.875%	15 April 2005	5,000,000,472
4.875%	15 April 2006	5,000,000,000
4%	15 January 2007	5,000,000,000
5.75%	15 February 2007	2,577,649,892
5%	15 April 2008	6,082,200,000
4%	15 April 2009	4,538,326,000
5.625%	15 October 2010	3,000,000,000
5.375%	15 October 2012	5,000,000,000

Source: EIB (2002).

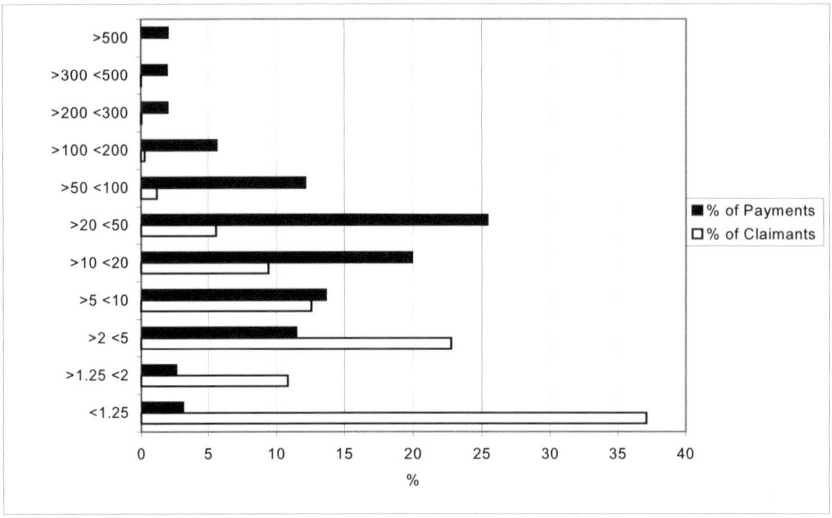

Fig. 7.1. IACS payments in 2000, categorised by payment size (in €1,000). Source: European Commission (2002a)

What would be the capital value of a bond, paying €10,000 in its first year? The answer depends upon the payment period, and the rate at which the market discounted the payments. But if there were ten annual payments of €10,000 in arrears, followed by €8,000 at the end of year 11, falling to €2,000 at the end of year 14, and a 4% discount rate applied, then the bond would be valued by the market at €93,610 at the time of issue. Clearly, with the passage of time (and particularly payments) its value would fall steeply, and it would have no residual value once the final payment had been made.

Calculated on the same basis, the total market capitalisation of all the bonds with an initial annual payment of €20.7 billion (the total IACS payment in 2000), when first issued, would be €194 billion, which is fairly large in comparison with the market values of the stocks listed in Table 7.2. However, the stocks listed in the table form only part of the whole euro bond market, and - in our view - creation of this new class of EU stock is unlikely to disrupt capital markets.

One factor that is likely to restrict the liquidity of the stock is that for practical and political reasons the registration of ownership/entitlement, and responsibility for annual payments, is likely to be retained by the Member State. To fully replicate the liquidity of the euro bond market, there would be a *single* authority. However, as existing IACS data would be required to implement the new system, and with IACS payments systems in place, it would be more pragmatic to continue to deploy this machinery. This would be the case whether a bond scheme had been introduced in place of, or as a successor to, the Single Payment Scheme proposed by the Commission for the mid-term review of Agenda 2000. The *shorter* the life of the bond, the more likely that these pragmatic considerations would apply. To some extent arbitrage operations might be expected to link these separate national sub-markets, but trading on the smaller markets would be thin. This suggests that if farmers want to sell their bonds they should do so in the early months of the scheme.

The retention of a national payments system would probably raise another issue: that of the payment rate outside Euroland (Denmark, Sweden and the UK at present; but in the new Member States too from Accession). Existing area and headage payments are, of course, denominated in euro, although in most instances paid in national currencies in the non-Euroland countries, and farmers face a risk as the exchange rate between the national currency and the euro varies. Under the proposed system, payments to bond holders would be in euro; but it would presumably be possible for a simplified system to apply, under which owners of a 'small' (say less than €1,000) annual payment entitlement could opt for payment in the national currency, at the prevailing exchange rate on the day of payment.

The budget

The proposal in Chapter 4 to convert existing arable area and headage payments (or, post-mid-term review, the new Single Payment) into compensation payments made to eligible bond holders would have a neutral impact on the EU's budget: in year one of the new scheme the same level of payments would be paid over to bond holders.

Administrative costs, borne by the Member States, should decline. Over time, as a result of degressivity, budget funds would be released for other uses, as discussed in Chapter 5. *If* compensation payments - area payments, an extension of the Single Payment Scheme, or new bonds - were extended subsequently to other sectors (such as sugar), budget costs would increase. But this follows from extending compensation to other sectors, not the adoption of a bond scheme *per se*.

Although budget neutral, year-on-year, in establishing payment obligations over a specified period (be it 5, 10 or 15 years) EU Ministers would be committing budget funds over a number of years, thus - potentially - exceeding their mandate. But, it could be argued, this commitment of budget funds has already occurred.

Expenditure on the CAP's price and income support mechanisms is often referred to as 'compulsory' or 'obligatory' expenditure, in that it flows automatically from CAP provisions (Harris *et al.*, 1983). Budget expenditure is the consequence of the policy; the budget allocation does not determine spend on the policy. Consequently the European Parliament's freedom to influence CAP budget allocations has been limited. Although few people expect that current payment schemes will last forever, the way that existing legislation is crafted means that the schemes would apply in perpetuity, unless amended or revoked. Thus, a *de facto* commitment of future budget funds has occurred that tends to crowd out other policy initiatives (Swinbank, 2002).

In the Agenda 2000 package agreed in Berlin in March 1999, the European Council placed a limit on budget expenditure on the CAP over the period 2000 to 2006, and the European Council in Brussels in October 2002 decided to extend the financial framework to 2013 (European Council, 2002). Whilst it is true that the EU has not committed itself to spend this money (instead the commitment is not to exceed these limits), the strong expectation is that these funds will be devoted to the CAP. Thus, the future financial commitments that the EU would undertake with the adoption of a bond scheme would not be too dissimilar to those already in place. A scheme lasting five or ten years would fall within the time scale of the latest 'framework of financial stability' extending to 2013.

Taxation

Member States retain the right to determine tax rates, and many treat agricultural businesses more favourably than other earning opportunities. Thus the profits of farm businesses, including the receipts from area and headage payments, can be treated quite differently from one Member State to another. The transformation of area and headage payments into a bond scheme would not change this.

However, the introduction of a bond scheme could trigger a change in the tax treatment of this income flow *within* a Member State, and thus render the proposed change less attractive to farmers. In the UK, for example, whilst area and headage payments are treated as trading receipts of the farm business, thus boosting the taxable trading profits of the farm business, it is not clear that bond payments would be treated in the same way[1]. Thus, even if received by the original farmer who remained on the land, it is probable that this bond income would be treated as unearned income. Consequently it would not qualify for tax relief on pension contributions, and it could not offset losses on the farming activities. Similarly, for capital gains tax purposes, and for inheritance tax, it is unlikely that the bond would be treated as an agricultural asset.

Any Member State that contemplates changing its tax legislation to treat the income stream received (and assets held) by farmers pre- and post-bond scheme in the same way faces a dilemma. If it legislates to ensure that bond payments received by *bona fide* farmers are treated as *farm* income, whilst bond payments received by private investors, and individuals who no longer farm but have retained their bonds, are treated as *investment* income, then there will be a tax inducement to retain the bond on the farm, and to continue farming.

On the other hand, an attempt to extend agricultural tax concessions to all bond holders, whomsoever they might be, would introduce complexities into the tax system, and set a precedent that most tax authorities would probably wish to avoid. Thus we conclude that this outcome is unlikely.

The most probable outcome is that the introduction of a bond scheme would not prompt Member States to change their tax legislation, and that in consequence some farm businesses would suffer a deterioration in their favourable tax treatment compared to the present IACS (or Single Payment) scheme. The farm lobby might, however, succeed in a campaign to change tax laws so that bond payments to practising farmers continue to be treated as receipts to the farm business.

Impact on production

In Chapter 4 we suggested that the introduction of a bond scheme, replacing area and headage payments, would return to farmers their freedom to farm. No longer would they have to keep animals, or grow crops, to qualify for IACS payments; instead payments under the bond scheme would be completely decoupled from production. This inevitably leads to questions about the likely response of farm businesses: will a switch in production take place, for example from cereal production into field vegetables; will farm land be used for some other purpose, or abandoned; and what will be the impact of these changes on the positive ('multifunctional') and negative externalities that are associated with agricultural production?

The European Commission's proposal, in the mid-term review, to replace existing direct payments with a Single Payment Scheme, raised similar questions; although the effect would probably be somewhat less pronounced as some crop specific payments would remain, and cross compliance - in particular the requirement that 'agricultural land throughout the EU be maintained in good agricultural condition' - suggests that, in reality, payments will not be wholly unconditional (Commission, 2002).

The Commission, and Ministries of Agriculture throughout the EU, struggled to quantify the likely impact of these proposals. Similarly, in our review of the literature on the bond scheme we have not found any robust estimates of the likely effects of the scheme. The problem is that the policy change proposed does not lend itself easily to modelling exercises based upon known parameters of farmers' response. Existing own and cross price elasticities of supply, for example, laboriously estimated by econometric techniques, do not tell us a great deal about how farmers would respond to a removal of the requirement to plant crops, or keep animals, whilst their revenue flows are maintained. Thus, whilst a modelling exercise would be a perfectly valid approach, the methodology adopted in this study was to undertake a major survey of how farmers thought they would react under certain circumstances, as reported in the next chapter.

Estimates vary considerably. In 1995, referring to a more radical policy change that would reduce all support prices to world market levels, and introduce decoupled compensation payments, economists at the British Ministry of Agriculture, Fisheries and Food remarked that agricultural production would fall, that 'the extent is difficult to predict', but that it 'seems likely the main effect would be a much more extensive farming rather than a major reduction in agricultural area'. However, 'elsewhere in the EU, where land had no alternative use, there would be a greater tendency for land to become derelict' (MAFF, 1995). By contrast, a working party of the National Farmers' Union for England and Wales suggested that 'large areas of more marginal land, e.g. in the LFAs, would be left un-farmed and allowed to become derelict' (NFU, 1994).

The Commission's proposals in the mid-term review prompted a number of studies and comments, some of which are noted below. For example, analysts from the British Meat and Livestock Commission suggested in October 2002 that the sheep flock in the UK could fall from 17 to 12 million head, and the suckler cow herd from 1.5 to 1.1 million head by 2010 (Agra Europe, 2002). In January 2003 the Irish Agriculture and Food Development Authority, Teagasc, published its results of a formal modelling exercise of the *decoupling* component of the July 2002 proposals (Teagasc, 2003). This suggested that the impact (by 2010) on EU-15 cereal production would be slight, whereas suckler cow numbers could fall by 18% and beef production by 6% (see Table 7.3).

The Commission's own impact analysis of the January 2003 proposals, released in March 2003, and the UK government's assessment of the same package, are also summarised in Table 7.3. It is perhaps important to stress that the Teagasc study encompassed only the decoupling component of the July 2002 proposals, whereas the other two embrace the full January 2003 package. One common element is the prediction that the fall in beef production, given existing import tariffs and other support mechanisms, will result in a marked increase in beef prices: thus the Commission's 2.7% fall in beef production by 2009/10 would lead to a 7.1% increase in beef prices, compared with its base-line prediction (European Commission, 2003).

Table 7.3. Alternative estimates of the impact of the mid-term review

	Commission, 2009/10	DEFRA	Teagasc, 2010
Scenario, compared to base-line predictions of existing policy package:	*January 2003 package*	*January 2003 package*	*July 2002 proposals for decoupling*
Cereal area			–2% (wheat) –1% (barley)
Cereal production	–2.1%	–5 to –10%	–1% (wheat) & 0% (barley)
Beef (suckler) cow numbers	–13.7%		–18%
Beef production	–2.7%	–5 to –10%	–6%
Ewe numbers			–7%
Sheepmeat production		–5 to –10%	–8%

Source: European Commission (2003); DEFRA (2003); Teagasc (2003)

In assessing likely response to decoupling, a number of considerations need to be borne in mind. First, that there may well be some delay in making change. One advantage of the bond scheme is that it maintains the revenue flow to farmers, thus avoiding the bankruptcies that would inevitably follow from an uncompensated policy change. Although economists predict that rational economic actors will equate marginal costs with marginal revenues, and produce accordingly, with the cushion of bond scheme payments it may take a number of years for all farms to make the adjustment. The pressure of the bank manager, of a new tenancy, or of a son or daughter hungry to take over the running of the business from the parent, is more likely to trigger change.

Second, alternative uses for the land may be limited. With milk and sugar production controlled by quota, and planting rights for vineyards severely constrained, a switch to these alternative enterprises is unlikely. Good agricultural land on the urban fringe tends to remain in agricultural production not because it is in its most profitable use, but rather that alternative uses are ruled out by planning constraints. A switch out of agriculture into non-agricultural use would, in a number of Member States, result in the loss of favourable tax treatment. In the LFAs, payment of compensatory allowances remains conditional on the continuation of agricultural activity; and, as suggested in Chapter 5, we would see the bond scheme embedded in a more fundamental reform of the CAP that would seek to support those multifunctional features of rural land use that society at large is willing to fund.

Acceptability to farmers

Another of the issues that we specifically addressed in the survey of farmers, reported in Chapter 8, is their perception of the acceptability of being in receipt of bond payments. The response to that question will be reported there. In this chapter we explore some other characteristics of farming in the Member States, and form some tentative conclusions about the likely variance in acceptability between countries.

One complaint about the existing pattern of IACS payments is that they are very unevenly spread, not only between Member States but also within Member States. A criticism of the new Single Payment Scheme, particularly from a Portuguese perspective, is that it will perpetuate, indefinitely, these inequalities (Cunha, 2002), but the regionalised scheme will mitigate this effect within Member States. Although a bond scheme, based on *compensation* for policy change rather than on *income* support, and of a finite duration, does address some of these concerns, they are not entirely dismissed. Table 7.4 illustrates another dimension.

Table 7.4. Numbers of holdings and IACS claims in the EU

	Holdings, 1997 (000)	IACS Claims, 2000 (000)
Belgium	67	44.50
Denmark	63	62.24
Germany	534	362.42
Greece	821	270.46
Spain	1,208	471.38
France	680	446.11
Ireland	148	129.06
Italy	2,315	691.65
Luxembourg	3	2.13
Netherlands	108	63.46
Austria	210	137.76
Portugal	417	174.02
Finland	91	71.52
Sweden	90	61.70
UK	233	146.35
EU-15	6,988	3,134.76

Source: total number of agricultural holdings, 1997 (European Commission, 2002b); number of IACS claims, 2000 (European Commission, 2002a)

Column two of Table 7.4 reports on the just under 7 million holdings in the EU in 1997, and column three on the around 3 million IACS payments made in 2000. There are two reasons why the numbers in column two might be higher than those in column three. First, on a number of farms, IACS claims will not be made (e.g. on fruit and vegetable farms, particularly in the Mediterranean), and, second, in a number of countries (e.g. the UK) the number of holdings may overstate the number of farms. One farm business may consist of several holdings as recorded in the national statistics. This does not just apply to the amalgamation of contiguous holdings. In the UK the authorities are strict in insisting that only one IACS claim can be submitted for any farms, wherever located in the UK, under common management.

What is particularly interesting is the outcome for Denmark. The evidence suggests that the bond scheme is particularly suited to Denmark because:

(i) most farmers would receive payments[2],
(ii) based on 2000 IACS claims, 75% of recipients would receive annual payments of between €2,000 and €50,000 (although 1.6% of recipients, in the group €50,000 to €100,000, would receive 10% of all the cash) (European Commission, 2002a), and
(iii) 75% of the land is owner-occupied, as we saw in Table 7.1.

We conclude that a bond scheme could, potentially, be 'sold' to Danish farmers on the premise that they are all receiving roughly equivalent treatment. By contrast, Table 7.4 suggests that only 42% of Portuguese farmers/holdings would receive a bond, and 80% of recipients would be entitled to small annual sums of €1,250 or less (European Commission, 2002a). Under these circumstances, it would probably be much more difficult to convince Portuguese farmers of the merits of a bond scheme. Although our survey only extended to three Member States, in Chapter 8 we are able to contrast the acceptability to farmers in Portugal, Germany and the UK of the proposed bond scheme.

Notes

[1] We are grateful to Adrian Baird for providing us with information from which we have constructed this paragraph. However, he bears no responsibility for the comments we make, or the inferences we draw.

[2] However, the number of IACS claims for Denmark seems suspiciously high. The number of holdings fell from 74,000 in 1993 to 63,000 in 1997 (European Commission, 2002b) and this trend will have continued.

References

Agra Europe (2002) 11 October, M/7.

Barthélemy, D. and David, J. (eds) (2001) *Production Rights in European Agriculture.* Elsevier, London.

Commission of the European Communities (2002) *Communication From the Commission to the Council and the European Parliament. Mid-term Review of the Common Agricultural Policy.* COM(2002)394. CEC, Brussels.

Commission of the European Communities (2003) *Proposal For a Council Regulation Establishing Common Rules for Direct Support Under the Common Agricultural Policy and Support Schemes for Producers of Certain Crops...* COM(2003)23. CEC, Brussels.

Cunha, A. (2002) A Mediterranean perspective of the CAP mid-term review. Paper prepared for Agra Europe's conference *Moving Towards Sustainable Agriculture? A Comprehensive Analysis of the CAP Mid-term Review.* 28 October 2002, Brussels.

Department for Environment, Food and Rural Affairs (2003) *Assessment of the Economic Impact of the Commission's Long Term Perspective for Sustainable Agriculture.* Agricultural Policy and Food Chain Economics Division, DEFRA, London.

Deutsche Bundesbank (2002) http://www.bundesbank.de/kredit_stripping.en.php

European Commission (2002a) *Commission Publishes Indicative Figures on the Distribution of Direct Farm Aid.* MEMO/02/198, 1 October 2002. European Commission, Brussels.

European Commission (2002b) *European Agriculture Entering the 21st Century.* Directorate-General for Agriculture, European Commission, Brussels (http://europa.eu.int/comm/agriculture/publi/reports/21century/index_eu.htm).

European Commission (2003) *Reform of the Common Agricultural Policy. A Long-term Perspective for Sustainable Agriculture. Impact Analysis.* Directorate-General for Agriculture, European Commission, Brussels.

European Council (2002) *Presidency Conclusions. Brussels European Council 24 and 25 October 2002.* Press Release. Council Secretariat, Brussels.

European Investment Bank (2002) Prospectus Supplement. *Euro Area Reference Note Issuance Facility Series No. 1583/01.* EIB, Luxembourg.

Frandsen, S.E., Gersfelt, B. and Jensen, H.G. (2002) Decoupling Support in Agriculture. Impacts of redesigning European Agricultural Support. Paper prepared for the *Fifth Annual Conference on Global Economic Analysis.* Taipei, Taiwan.

Harris, S., Swinbank, A. and Wilkinson, G. (1983) *The Food and Farm Policies of the European Community.* John Wiley & Sons, Chichester, UK.

House of Commons (2002) *Agriculture Fraud: the Case of Joseph Bowden.* Public Accounts Committee. 51st Report Session 2001-02, HC 684. The Stationery Office, London.

MAFF (1995) *European Agriculture: the Case for Radical Reform. Working Papers.* Prepared by the Economics (International) Division for MAFF's CAP Review Group. MAFF, London.

Marsh, J.S. *et al.* (1990) *A Future for Europe's Farmers and the Countryside.* Land Use and Food Policy Inter-Group, European Parliament.

National Farmers' Union (1994) Long Term Strategy Group. *Real Choices.* NFU, London.

Ravenscroft, N., Gibbard, R. and Markwell, S. (1999) *Private Sector Agricultural Tenancy Arrangements in Europe - a Critical Review of Current Literature.* Working Paper No 28. Land Tenure Center, Madison: University of Wisconsin, Madison, Wisconsin.

Rodgers, C.P. (1998) *Agricultural Law.* 2nd edition. Butterworths, London.

Swinbank, A. (2002) Direct payments and the impasse over CAP Reform. *Agra Europe* No 1991. 22 February 2002, A/1-A/27.

Teagasc (2003) *An Analysis of the Effects of Decoupling Direct Payments from Production in the Beef, Sheep and Cereal Sectors*. Report prepared for the Department of Agriculture and Food by the Rural Economy Research Centre. Teagasc, Dublin.

Chapter Eight:

Asking Farmers about their Response to the Proposed Bond Scheme

Richard Tranter[1], Leonardo Costa[2], Thomas Knapp[3], Jonathan Little[1] and Miguel Sottomayor[2]
Centre for Agricultural Strategy[1], The University of Reading, UK; Faculty of Economics and Management[2], The Portuguese Catholic University, Portugal; Institut für Agrarökonomie[3], Georg-August Universität Göttingen, Germany

Introduction

The aim of the postal survey and farmer interview phases of the research project was to establish the effects of farmers' stated responses, or intentions, to the proposed bond scheme on farm business structure, rural employment, output and land use in the UK, Germany and Portugal. Through this, the practicability and acceptability of our proposed scheme were to be assessed. Furthermore, although this was not the original idea behind the research project, this phase of the project might also be useful in assessing how farmers might respond to the new Single Payment Scheme.

Asking farmers about the future

Before presenting and discussing the findings from the postal survey and farmer interviews, it is worth reviewing some work that has been done on asking farmers about the future to set these in context. For some 40 years, studies of farmers' future plans and intentions have been carried out from time to time for two main reasons:

1. In order to help discover what information and knowledge aids farmers in making decisions about the future and, indeed, how they then make decisions.
2. As a way of obtaining information for both public and private policy-makers on likely future developments in farming.

Examples of intentions surveys in agriculture are provided by Johnson and Haver (1960) and Nielson (1962) for a wide range of farms in the mid-West USA; by Reithmuller (1978) and Munro and Fisher (1982) for graziers in Australia; and by Pryde (1982) and Pryde and McCartin (1983) for a broadly based sample of farmers in New Zealand.

Surprisingly little direct work on intentions has been done in the UK, though the work of Thornton (1961) on the business plans of pig producers in South Central England is a seminal work. Thomson and Tansey (1982) reviewed various surveys of dairy farmers' future intentions and Poole (1983) and Gasson (1983) discuss related aspects from surveys of small milk producers and part-time farmers in England respectively. Jones *et al.* (1987) asked farmers in England how they would structure the use of information in making various decisions about the future of their businesses and Fearne (1990) reported on a survey of the reasons why farmers might change their practices in the future.

More recently, Gasson *et al.* (1998) examined how English farmers were planning to adapt to the twin pressures of falling farm incomes and arrangements for retirement, inheritance and succession, and Holt *et al.* (2001) examined the reasons behind planned structural changes on farms in Central Southern England. ADAS (2002), in a project examining how government could better deliver advice to farmers, asked a sample of farmers in England and Wales what topics they would be seeking information on in the next 12 months. Perhaps the most comprehensive farmers' intentions survey in recent years was reported by Harvey (2000). It consisted of monitoring estimates of investment intentions, determining perceptions of the agricultural business environment and output, and obtained details of resource allocation plans from all the farms in the Farm Business Survey in England and Wales.

There are, of course, problems in both carrying out intentions surveys and interpreting their results. Designing data collection instruments and questionnaires is always difficult and care has to be taken not to prompt the farmer into a response that they do not really believe or mean. Additionally, younger and better-educated farmers may respond more positively to requests to take part in such surveys and they may be more used to, and feel more comfortable with, answering questions about likely future actions than other farmers. Questionnaires can also become long or wordy as investigators make attempts to define the conditions under which farmers' intentions are to be expressed.

Perhaps the most crucial question concerning farmers' intentions surveys that needs addressing is whether they provide accurate answers. In other words, do farmers actually behave in the future in the way they say they will at the time when they were questioned? Very few studies have examined this issue. However, Thomson and Tansey (1982), Gasson *et al.* (1998) and Harvey (2000) did do this and found that, in their studies, most farmers did actually do what they said they would, especially for relatively short-term decisions or actions. If this accuracy applies across the board, then the alternative of using prediction and forecasting models such as those reviewed by Bauer (1989), Harvey (1990) and Johnson and Rausser (1977) and which involve making a series of assumptions about social, economic and technical issues, seems a less appealing prospect for those seeking how farmers might act in the future. For this reason then, for the research project discussed here, it was decided that the intentions survey approach, backed up by full stakeholder consultation in person, by post and through workshops was the most appropriate approach to adopt.

Methodology

Following a pre-pilot phase, and a formal pilot survey exercise of 150 farmers, the postal survey was carried out in each study country during the autumn and winter of 2001/02. This survey was closed after 20 weeks and a series of 150 on-farm interviews, with farmers drawn at random from respondents to the postal survey, followed in the spring of 2002.

In the UK, farm incomes at that time were recovering slightly from their low point in 2000, but the farm sector had been very badly shaken by the outbreak of foot and mouth disease that had raged through most of 2001. Indeed we had delayed our survey because of the outbreak (and, earlier, because of the BSE crisis that had erupted in Germany in the autumn of 2000), and only embarked upon it once the disease was contained. The postal survey, and the following on-farm interviews, however, predated the launch of the mid-term review of the CAP in July 2002.

The survey sample used was around 4,500 farmers in each of the three countries. This was drawn from the 'Yellow Pages' telephone book for the UK, the Pensions records for Germany and the list of the Office of National Statistics for Portugal. An expensively produced four-page questionnaire[1] was used which included a stepwise approach to defining the new policy scenario (the proposed bond scheme). A personalised 'covering' letter accompanied it and reminders were sent out after four weeks and eight weeks. The questionnaire design and the procedure employed were as identical as possible in each country. Reminder letters stressed how we wanted broad coverage and how, by responding, farmers would themselves be making a personal contribution to the framing of future agricultural policy. All respondents were thanked by post.

The crude response rate, shown in Table 8.1, was 38.6% for the UK, 31.8% for Germany and 32.3% for Portugal. However, if allowance is made for people who were no longer in farming by subtracting these from the original total despatched, the 'real' response rate is computed as 40.2% for the UK, 36.8% for Germany and 33.4% for Portugal. These rates are especially high for voluntary postal surveys of farmers.

Clearly it was important to check for bias in the sample of respondents before examining and analysing the replies. So, when respondents' characteristics were compared with the known overall national pattern, it was found that the survey response might under-represent smaller farm businesses in both the UK and Portugal. However, an investigation into non-response bias, by testing the responses of the first third against the last third, found very few statistically significant[2] different features. Using this procedure, if the last third were significantly different from the first third, then it is usually taken that those who did not reply at all are more likely to be similar to the last third of respondents. For example, in the UK, non-respondents were likely to be less well-educated than respondents (at the *** level), while in Portugal, those who responded later were significantly younger than those who did not return their questionnaires at all (**). In Portugal, this finding was also proved by the fact that because the sample was drawn from the official national census database, we knew certain characteristics of the sample in advance, such as age.

Table 8.1. Survey response rates by category of reply

	UK		Germany		Portugal	
	No.	%	No.	%	No.	%
Total questionnaires despatched	4,499	100.0	4,500	100.0	4,517	100.0
Total replies received of which:	1,970	43.8	2,160	48.0	1,820	40.3
Letters returned stating addressee 'not known', 'gone away', 'deceased, 'non-farmers' or 'retired'	176	3.9	621	13.8	151	3.3
Refusals or unusable replies	58	1.3	110	2.4	212	4.7
Completed questionnaires	1,736	38.6	1,429	31.8	1,457	32.3
'Real' response rate		40.2		36.8		33.4

The series of 150 on-farm interviews in the three countries had a dual purpose. First, they acted as a form of 'validation' to the postal survey answers and, second, they allowed us to gain a more detailed insight into the reasons behind the respondents' answers to the various questions on the questionnaire. These case-study visits were chosen to roughly represent the numbers of farmers from each part of each country that had responded to the questionnaire. The other selection criteria used to stratify the sample for visits included the predictions given for change in response to the policy scenario, the age of the farmer and the size of the farm. Whilst a common standard interview schedule of questions was used in each country, respondents were given several open-ended questions to enable them to discuss matters in a broad sense. Many thousands of kilometres were covered in each country by the interviewers travelling between the farms.

Farmers' views on the future of farming

Before describing the proposed changed policy scenario, and asking for respondents' reactions to it, we asked for their level of agreement with seven statements on the future of farming. This was done for two reasons. First, to get the farmers introduced to the idea of being asked questions about the future and, second, so that we could see whether there was any link between their views on the future and their reaction to the changed policy scenario - the bond scheme.

The farmers were asked to show their strength of agreement with the statements using a five-point scale (five representing 'strongly disagree', one 'strongly agree', whilst three indicated 'uncertain'). Table 8.2 shows the mean 'scores' for each statement listed country by country. Interestingly, there was much similarity between the answers for each of the three study countries as the statements which received the highest overall mean scores (or highest agreement

Table 8.2. Respondents' views on statements on the future of farming

		Overall mean score		
	Statements	UK	Germany	Portugal
1.	The future of farming is dependent on continued support	2.01	2.01	2.07
2.	In my opinion the current arrangements under the CAP work well	3.17	3.84	3.29
3.	The existing Agenda 2000 arrangements will be continued after 2006	2.97	2.51	2.80
4.	Due to pressures for change, the current system will need to be reformed	2.13	2.33	2.42
5.	Current direct payments are a reliable source of income for farmers	1.91	1.59	2.28
6.	If current support is withdrawn, many farms would become unprofitable	1.55	1.61	1.83
7.	Support for agricultural production should be phased out gradually	3.11	3.62	3.82

Key:
1 = 'Strongly agree'
5 = 'Strongly disagree'

level) were: 'If current support is withdrawn many farms would become unprofitable' followed by 'Current direct payments are a reliable source of income for farmers' and 'The future of farming is dependent on continued support'. The statement that provoked most disagreement amongst our respondents was 'In my opinion the current arrangements under the CAP work well', reflecting a general dissatisfaction with the status quo.

In an attempt to discover reasons behind farmers' views on the future of farming, we looked for differences in agreement levels on the seven statements between six sub-divisions of farm and farmer type - size and type of farm for example. Many statistically significant differences were found. However, if the highest scored statement is concentrated on, 'If current support is withdrawn, many farms would become unprofitable', then it was found that, in the UK, more of those with farms of over 100 ha agreed with this than those with smaller farms (***) and those with livestock farms agreed less than those with mixed or mainly cropping farms (***). In Germany, those who left full-time education before they were 20 were more likely to agree with this statement than those who left later (**). In Portugal, those with farms of 25 ha or over were more likely to agree with the statement than those with smaller farms (*); this is probably due to the fact that the larger farms in that country tend to be cereal or livestock farms and hence have direct support under the CAP, whereas the smaller farms, which tend to be producing fruit, vines and vegetables, do not.

The statement that was the second highest agreed with by farmers in the three countries was 'Current direct payments are a reliable source of income for farmers'. In Germany and Portugal there were no significant differences for this statement by farm or farmer type. However, in the UK, there were three such differences: livestock farmers were less strongly in agreement with this statement than were farmers with crops (**); farmers aged up to 50 agreed less with it than older ones (***); and farmers who left full-time education before they were 20 agreed with it much more strongly than those who left it later on in life (***).

Farmers' reactions to payments being decoupled from current land use and production

The proposed changed policy scenario - bond scheme - was introduced to our respondents gradually on the questionnaire. Thus, it is possible to make some inferences regarding the production impacts of the proposed policy change from the farmers' reactions to the decoupling of payments from current land use - the first two steps of the bond scheme as outlined in Chapter 4 above. This scenario corresponds, more or less, to the Single Payment Scheme subsequently adopted in the Fischler reforms. Table 8.3 shows that around 67-69% of the respondents said they would *not* alter their mix of farm activities after the first elements of the proposed policy change were introduced. It is interesting to see how close this proportion was for each of the three countries, perhaps surprisingly so. Table 8.3

also presents answers by sub-groups of our respondents - for total area farmed, type of farm, age of respondent and the age the respondent left full-time education. These, as can be seen, were examined for significant differences. It was found that, in the UK, more of those with larger farms (***), those that were mixed (**) and those who were aged up to 50 (***) or who had had some level of higher education (*) said they would alter their mix of farm activities. For Germany, there were no statistically significant differences by these sub-groups and, for Portugal, only one - that those with larger farms were more likely to alter their activity mix (*).

Table 8.3. The proportion of respondents who would alter their mix of activities under the two decoupling steps of the proposed bond scheme by type of farm and farmer

	Proportion (%) of respondents who:					
	UK		Germany		Portugal	
Characteristics of farm/farmer	Would not alter	Would alter	Would not alter	Would alter	Would not alter	Would alter
All farmers	69.1	30.9	66.8	33.2	67.1	32.9
Total area farmed[1]:						
<100 ha	73.1	26.9	69.8	30.2	69.6	30.4
100 ha and over	64.9	35.1	50.5	49.5	59.7	40.3
	χ^2=13.251 df 1 ***		χ^2=3.198 df 1 ns		χ^2=4.786 df 1 *	
Type of farm:						
Mainly livestock	71.5	28.5	66.2	33.8	70.9	29.1
Mainly cropping	70.1	29.9	69.2	30.8	67.5	32.5
Mixed	63.6	36.4	62.0	38.0	60.9	39.1
	χ^2=9.516 df 2 **		χ^2=3.784 df 2 ns		χ^2=5.322 df 2 ns	
Age of farmer:						
Up to 50	64.3	35.7	67.8	32.2	64.3	35.7
51 and over	72.4	27.6	64.7	35.3	67.9	32.1
	χ^2=12.641 df 1 ***		χ^2=2.387 df 1 ns		χ^2=1.344 df 1 ns	
Age at leaving full-time education:						
Up to 19	70.4	29.6	66.9	33.1	67.8	32.2
20 and over	63.9	36.1	62.7	37.3	64.3	35.7
	χ^2=5.021 df 1 *		χ^2=0.983 df 1 ns		χ^2=0.568 df 1 ns	

[1] In Portugal, less than 25 ha, and 25 ha and over.

Those respondents who said they would alter their activity mix (some 32% of all those who replied), as a result of the introduction of the first two steps of the bond scheme, were then asked a follow-up question regarding the nature of their likely change in activity or enterprise. This was not only to obtain an assessment of the production impact the suggested policy change would have on the existing farm enterprises, but also to gauge the potential extent of farmers switching between enterprises, with full-decoupling now freeing them to move between sectors without the fear of losing any of their direct payments.

Table 8.4 shows the nature of the three countries' respondents' proposed change in their *main* activity. It includes all respondents together, classified by main farming activity, including those who had answered they would not alter their mix of activities. The overall picture of a reduction in enterprises involving sheep and beef cattle, as a main activity, is most likely due to the breaking of the link between the amount of headage payments and livestock numbers. The predicted increases in dairying, and the results for cereals, may well reflect these farmers' belief that it will be necessary to expand the scale of their main activities in order to spread costs and become more competitive under any policy reform scenario. The general tendency for increases in other activities, such as permanent crops, pigs and poultry, horticulture and other vegetable and root crop production under the category of general cropping, suggests that farmers perceive the possibility of an upturn in the relative profitability of these, currently less supported, sectors under the proposed policy change. The right-hand column in the table, though, serves as a reminder that, as we saw in Table 8.3, the large majority of the respondents in each country indicated that the suggested policy reform was unlikely to prompt them to make any changes at all to their current mix of farm activities. This finding could well provide comfort to policy makers.

Table 8.4. The nature of changes in main enterprise under decoupling within farm type categories, UK, Germany and Portugal together

	Nature of change in activity (% of total)		
Activity	Increase	Decrease	No change
Dairying	19.3	5.2	68.1
Extensive sheep / cattle	6.8	14.0	69.8
Intensive sheep / cattle	6.9	9.9	70.7
Cereals	10.2	10.0	64.4
General cropping	9.9	3.2	65.4
Permanent crops	10.4	3.8	67.9
Pigs / poultry	10.9	3.6	59.9
Horticulture	12.2	4.1	62.2

The on-farm interviews enabled further investigation of respondents' answers to this question which, because of the necessary simplicity of the postal questionnaire, clouded their real views. For example, it appeared from the interviews that some of the stated changes to the cereals activity in fact related to respondents wanting to drop one cereal crop and replace it with another in the arable rotation. It also turned out that some of the responses relating to an 'increase' in activity really meant that they felt they had to expand their business overall by taking on more land in order to maintain future profitability, rather than increasing the intensity of production on their existing area.

Looking at Table 8.4, one might wonder why the proportions across the columns do *not* sum to 100. This is because the changes shown in the first two columns are only for the *main* activity on the farm; the remainder represents predicted changes to their secondary activities. As a result, it is the analysis of changes to these secondary activities, as well as the adoption of new activities currently not present in the farming mix, which gives the best guide to the amount of switching between enterprises that will result from the suggested policy change. The fact that over twice the number of cereal farmers said they would increase, rather than decrease, activities such as permanent crops, horticulture and general cropping, might be thought to support fears that decoupling would have a destabilising effect on fruit and vegetable markets. However, it should be noted that the numbers of our respondents predicting such a change are relatively small, representing only around 7% of all the mainly cereal farmers in our total sample.

Table 8.5 shows the proportion of respondents predicting the adoption of a new activity. In the three-country sample, there are almost 550 instances of an activity being adopted as new, but this did not mean that 550 farmers predicted they would adopt a new activity, as respondents could, and indeed did, indicate the likelihood of adopting more than one new activity. It seems from these results that the proposed new adoption of activities in numbers is pretty evenly spread, apart from more for non-food crops and forestry and woodland. However, this weighting is slightly biased by the Portuguese results, with considerably more of their farmers anticipating a move into an alternative activity, with forestry being the most popular choice.

Table 8.5. The proportion of respondents indicating they would adopt a new activity under decoupling, the UK, Germany and Portugal together

Activity	Proportion of respondents (%)
Dairying	7.1
Extensive sheep / cattle	9.0
Intensive sheep / cattle	9.5
Cereals	7.0
General cropping	8.2
Permanent crops	7.5
Pigs / poultry	7.5
Horticulture	9.2
Forestry / woodland	14.1
Non-food crops	12.8
Other activities	8.1

Another fear of the introduction of the bond scheme, which emerged from the consultations with stakeholders, concerned the expectation that large areas of the countryside would be left idle with all the associated problems of landscape damage, rural unemployment, fire risk from an increase in scrub and the possible loss of biodiversity. Therefore, in order to explore this issue, the surveyed farmers were asked whether they would leave any of their land idle under decoupling. Their answers to this question are summarised in Table 8.6, where it can be seen that at least 50% of those in each of the three countries would leave no land idle at all - in the UK this figure is around 80%. And, considerable numbers also would idle less than half their land. So, it seems from our results, that few of our respondents would make use of what is effectively a relaxation of the 50% maximum set-aside restriction, to eliminate all of their fixed costs, although there were country differences. For example, the Portuguese farmers were much more likely than the German or UK farmers to do this by idling all their land.

Within these overall figures, there were various interesting statistically significant differences according to farm and farmer type. For example, in the UK, relatively more of the respondents with 100 ha or over would leave land idle than those with less (***) whereas, in Portugal, relatively more of those with less than 50 ha farms would leave land idle than those with larger farms (*). In the UK, farmers who had 50% or more of their income from non-farm sources were more likely to leave some land idle than those who obtained the majority of their income from the farm (***). In Portugal, considerably fewer of the respondents who said a successor was definitely or very likely to have been identified would leave land idle, compared to those who had not identified a successor (***).

Table 8.6. Respondents' answers to the question of whether they would leave any of their land idle under decoupling

	Proportion of respondents (%)		
Action	UK	Germany	Portugal
None	79.9	59.7	52.2
Less than half	15.4	28.7	24.4
Around half	3.1	4.7	5.6
More than half	0.7	2.5	8.5
All	1.0	4.4	9.3

Some concern arose in the stakeholder consultation phase of the project that decoupling would lead to a fall in the level of agricultural production and hence affect the whole rural economy, especially the ancillary industries. At the other end of the spectrum, fears were also expressed that an increase in intensity would damage the environment. Thus, we asked a question to try to answer these fears. Table 8.7 shows that around 50-60% of the respondents in each country would not change their level of intensity of production. However, nearly 20% each in Portugal and the UK would increase or decrease their intensity of production in response to the first steps of the bond scheme. However, in Germany, whilst few said they would increase production intensity, over a third said they would decrease it.

Table 8.7. Respondents' answers to the question of whether they would intensify production under decoupling

	Proportion of respondents (%)		
Action	UK	Germany	Portugal
Greatly decrease	2.2	10.6	7.9
Decrease	18.7	25.7	18.2
Remain unchanged	59.3	60.3	52.6
Increase	18.3	3.1	19.6
Greatly increase	1.5	0.3	1.6

Table 8.8 shows that, whilst just over 70% of the respondents in the UK would not change their level of labour use under the first steps of the bond scheme, in Germany and Portugal, the equivalent figure was some 55%. Relatively few in each country would increase it.

Table 8.8. Respondents' answers to the question of whether they would change the amount of labour employed under decoupling

	Proportion of respondents (%)		
Action	UK	Germany	Portugal
Greatly decrease	2.9	7.1	9.1
Decrease	20.6	32.3	21.6
Remain unchanged	72.7	55.6	55.1
Increase	3.7	4.4	13.2
Greatly increase	0.1	0.5	0.9

During the stakeholder consultation phase of the project, concern was expressed in each of the three study countries that public opinion would not look favourably on the bond scheme. The reason put forward, and it was most vocally expressed by politicians and officials of farmer organisations and unions, was that it might seem that farmers would be rewarded by society for 'doing nothing', i.e. they would not be required by law to farm in order to receive support. So, in order to shed light on this possibility, we asked our respondents 'How would you feel about receiving payments unattached to production decisions?'. In the UK, less than 25% of the respondents said they would feel uncomfortable at receiving payments unattached to production decisions under the first steps of the proposed bond scheme. This contrasted strongly with the Portuguese results, where almost 70% were 'quite' or 'very' uncomfortable at the thought of receiving 'fully' decoupled payments. The German position was somewhere between these two figures at 45% of the respondents feeling 'quite' or 'very' uncomfortable.

Farmers' reactions to payments being decoupled from the land and being attached to individuals

On the questionnaire, once the farmers' reactions to the first two steps of the bond scheme had been ascertained, the next step of the changed policy scenario was explained to them. It will be remembered, from Chapter 4, that this entailed decoupling payments from the land (or farm) and attaching entitlements in the form of a certificate to those farming the land at that time. It was also explained that this entitlement would be inheritable. As outlined in Chapter 4, Table 4.1, this step would have the benefit of stopping the distortion of land prices by the capitalisation of expected future payments and thus help farm restructuring. New entrants from then on would not expect to receive direct payments, but would expect to pay lower land prices to enter farming. Nevertheless, this benefit was not spelt out on our questionnaire.

Once the third step was explained, respondents were asked to show how strongly they agreed with four questions comparing their current situation with that which they thought likely once the third step of the bond scheme had been introduced. Table 8.9 shows the overall mean scores provided by all the respondents, country by country - a five-point scale was used with five representing 'strongly disagree' and one 'strongly agree'. Intriguingly, it will be seen that, in rank order terms, respondents in each of the countries agreed in almost the same way. For example, the statement that received the lowest overall mean score, implying the strongest agreement was, 'All payments should go to the tenant or operator rather than to the landlord', echoing sentiments expressed during the stakeholder consultation. Perhaps surprisingly, right across the three countries, there was little agreement that the introduction of the third step of the bond scheme would lead to a significant fall in land prices. And, the statement they failed to endorse most was, 'The freedom and flexibility provided by the change would benefit your business', suggesting that they did not hold a strong opinion about the greater liberty embodied in the policy scenario.

There were several statistically significant differences in agreement levels by farm and farmer types. Most noteworthy was the fact that, in Portugal, livestock farmers agreed more strongly with the statement that all payments should go to tenants/operators than mixed farmers and crop farmers (at the * level). In Germany, the same finding occurred (***) as it did in the UK (**). This seems to reflect the importance of this issue to tenant farmers in the livestock sector, who may well themselves have purchased beef and sheep quota and now fear that the value of their capital asset would unreasonably be lost if the landlord was to be the recipient of the entitlement to future payments. On the same question of allocation of the entitlement to the tenant rather than the landlord, younger farmers agreed more strongly than those aged over 50 in Portugal and in Germany, both at the ** level of significance.

Table 8.9. Respondents' agreement with questions on the likely situation after decoupled payments are attached to individuals

	Questions	Overall mean score		
		UK	Germany	Portugal
1.	Land prices will decrease significantly	2.48	2.67	2.82
2.	All payments should go to the tenant or operator rather than to the landlord	1.98	1.71	2.43
3.	The change in capital values would have an impact on your ability to borrow	2.31	2.37	2.66
4.	The freedom and flexibility provided by the change would benefit your business	2.86	3.28	2.66

Key:
1 = 'Strongly agree'
5 = 'Strongly disagree'

Farmers' reactions to receiving the bond in lieu of the previous support payments

The latter part of the questionnaire was designed to assess farmers' reactions to receiving the bond in lieu of the previous support payments - steps four, five and six of the bond scheme (see Chapter 4). The questionnaire explained that the future level and duration of payments would be fixed - at the current levels for ten years and then tapering to zero after year 15. The farmer would be given a certificate, or bond, guaranteeing the stream of payments. Finally, it was detailed that the bond could mean that, if held, annual payments would continue for 15 years, or it could be sold on the capital market into a money sum that could be used for a variety of on- and off-farm purposes. As summarised in Table 4.1, these steps have the great benefit of establishing certainty into policy and locking-in policy reform.

As with the third step of the bond scheme discussed above, once steps four, five and six were explained on the questionnaire, respondents were asked to try to reflect their practical response to the proposed changes when answering a new series of questions.

A key indicator of farmers' future intentions came from their responses to the question asking what they might do with the bond. Their answers, summarised in Table 8.10, suggest a striking reluctance to cash in the bond on the financial market, with around half or more of the respondents indicating that it was 'unlikely' or they definitely would not do so. This response was particularly pronounced in Germany, where some 63% answered in this way. Whilst at least a quarter remained open to the possibility, only between 10 and 20% of respondents said they would 'definitely' or were 'very likely' to cash in their bond.

Table 8.10. Respondents' answers to the question of whether they would cash in their bond by selling it on the financial market

	Proportion of respondents (%)		
Action	UK	Germany	Portugal
Definitely	2.2	6.4	6.1
Very likely	9.5	7.0	14.3
Possibly	37.3	23.4	32.9
Unlikely	42.3	24.0	27.6
Definitely not	8.7	39.2	19.1

The understanding of the responses to the question on the sale of the bond, in particular, has improved as a result of the 50 on-farm interviews carried out in each country. During these face-to-face conversations, we gave each of the farmers an estimate of the market value of their own individual bond, based on their recent receipts of direct payments. While we did not speak to anyone with a sufficiently large enough farm to be affected by the Commission's original proposal to cap direct payments, we did still visit some potential 'bond millionaires' and it was the size of their potential asset that seemed to surprise many of the farmers we spoke to. While we had hoped that the respondents to the questionnaire would have grasped for themselves the approximate value of their own bond entitlement, some of the farmers interviewed in the UK and Germany seemed much more willing to consider the possibility of selling the bond than they had suggested in their questionnaire return, after the likely full value of their potential asset was spelt out to them. Many did, however, feel that they might be discouraged from capitalising their asset if the tax position left them seriously disadvantaged in comparison with those retaining the bond to its proposed expiry date in year 15. For instance, some thought that they would surrender the favourable capital taxation treatment that farmers have in each of the study countries.

Now, one possible benefit to society of the proposed bond scheme is that it could be used to finance some form of business diversification which might help the wider rural economy. When asked whether they would consider a diversification activity after being issued with the bond, Table 8.11 shows that only between 5 and 12% of the respondents said they 'definitely' would, and a further 12-15% said that it was 'very likely' they would. Considerable numbers said they 'possibly' would. Interestingly, more than half of our German farmers indicated a limited opportunity for diversification, again probably as they feared it would alter their relatively favourable taxation position. When statistically significant differences were examined, it was not surprising to find that in Portugal (*) and in the UK (**) farmers aged up to 50 were much more likely to say they would consider some form of diversification in response to actually receiving the bond. Furthermore, in the UK, those with farms with over 100 ha were more likely to do this than those with smaller farms (**).

Table 8.11. Respondents' answers to the question of whether they would consider some form of diversification of their farm business in response to the full bond scheme implementation

	Proportion of respondents (%)		
Action	UK	Germany	Portugal
Definitely	8.1	12.3	5.5
Very likely	15.2	15.4	12.5
Possibly	44.3	19.0	34.8
Unlikely	27.2	18.7	29.8
Definitely not	5.3	34.6	17.5

Possible impediments to diversification were discussed in each country as part of the series of on-farm interviews. Some of the reasons for not considering diversification included planning and tenure restrictions, and remoteness from urban populations.

Table 8.12 presents respondents' answers to the question of whether they thought the 15-year transition period, encapsulated in the bond scheme proposals, would be sufficient to enable them to adapt to the new policy environment. It stands out that many (between 44 and 59%) thought 15 years an 'adequate' period, with between 26 and 49% thinking it to be 'too long' or 'much too long'; relatively few (especially in Portugal) thought 15 years 'too short' or 'much too short'.

A few statistically significant differences between farm and farmer sub-groups were apparent. Most noteworthy was the finding that farmers with less than 100 ha were much more likely than those with larger farms to think 15 years as 'too long' or 'much too long' in the UK (***) and in Portugal (*). Despite findings reported earlier in this chapter that our respondents, on average, made little of the flexibility embodied by the bond scheme, it seems the majority of them felt they could adapt to the proposed changes relatively quickly.

Table 8.12. Respondents' answers to the question of whether a transition period of 15 years would be sufficient to adapt to the new policy environment

<table>
<tr><th></th><th colspan="3">Proportion of respondents (%)</th></tr>
<tr><th>Action</th><th>UK</th><th>Germany</th><th>Portugal</th></tr>
<tr><td>Much too long</td><td>6.3</td><td>12.4</td><td rowspan="2">48.5</td></tr>
<tr><td>Too long</td><td>20.0</td><td>20.1</td></tr>
<tr><td>Adequate</td><td>59.1</td><td>44.4</td><td>44.3</td></tr>
<tr><td>Too short</td><td>12.0</td><td>15.2</td><td rowspan="2">7.2</td></tr>
<tr><td>Much too short</td><td>2.6</td><td>7.9</td></tr>
</table>

Farmers' views on whether they still will be farming in ten years' time

At three points of the postal questionnaire, respondents were asked whether they thought they would still be in farming in ten years' time and, if not, what they thought they would be doing. These questions were posed to try to assess the structural impacts on the industry of the changed scenario and to test for any difference in reaction according to the sequential elements or steps of the proposed bond scheme. The question was first posed before the proposed changed policy scenario was outlined. It was then posed again after the full implications of decoupling had been explained. The third, and final, time the question was posed was after the respondents had been told about receiving the bond and what they could do with it.

It can be seen from Table 8.13 that the proportions who said they thought they would remain in farming after ten years vary markedly between the three countries. The relatively low proportion who said they would be continuing in Portugal may be due to the age structure of their industry, with over half those falling out of agriculture being set to retire at the normal retirement age. Interestingly, however, over two-fifths of those who said they would be leaving the industry in Portugal expected to take up other employment, a much higher proportion than was seen with the UK sample. During the subsequent interviews it became clear that 'other employment' tended to mean employment in urban and more developed regions both internally in Portugal and abroad. While there appears to be a general slightly declining trend in the numbers saying they would remain in the industry in all three countries, after the successive elements of the proposal were introduced to them, these differences were found not to be statistically significant in the UK and Portugal. However, in Germany, there was a significant difference (**). The main difference seen in the German results is, for the most part, due to those who had not answered the original question then being prompted into making a negative response when the question was posed a second time, following the explanation of the policy scenario.

Table 8.13. The proportion of respondents who thought they would still be farming in ten years' time, currently and after the staged introduction of the bond scheme

	Proportion of respondents (%)		
Time of Question	UK	Germany	Portugal
Now	52.3	61.3	41.5
Decoupling current payments from production	50.5	54.6	40.3
At receipt of the bond	49.2	55.7	39.7

Farmers' views on various statements on agricultural policy

The final part of the questionnaire investigated, in very general terms, farmers' views on some agricultural policy matters. Table 8.14 presents the mean 'scores' of the respondents, country by country, as to how strongly they agreed with the five different statements on agricultural policy options they were given (1

Table 8.14. Respondents' views on agricultural policy options

		Overall mean score		
	Statements	UK	Germany	Portugal
1.	Agricultural support should be tied to environmental benefits	2.80	2.76	2.22
2.	Payments to farmers should be modulated on the basis of farm size	2.65	2.36	1.73
3.	Payments to farmers should be related to regional disadvantage	2.23	1.93	1.79
4.	All sectors (including milk and sugar) should have current support converted into bonds	2.95	3.18	2.77
5.	Savings from our proposed bond scheme should be directed to environmental schemes	2.97	3.05	2.40

Key
1 = 'Strongly agree'
5 = 'Strongly disagree'

represents 'strongly agree' and 5 'strongly disagree'). It can be seen that, broadly, the two statements that farmers in all three countries were in strongest agreement with were: 'Payments to farmers should be related to regional disadvantage' and 'Payments to farmers should be modulated on the basis of farm size'. These feelings were at their strongest for the Portuguese farmers which, on reflection, should have been expected in view of the large number of small farms in that country as well as large areas of severely disadvantaged mountainous land. German and UK farmers did not express particularly strong views on the idea that savings from our proposed bond scheme should be directed towards environmental schemes. The most negative response of all was from the German farmers regarding the notion of support in other sectors (such as milk and sugar) being converted into bonds.

As one might expect, with much livestock production being concentrated in disadvantaged, often hilly areas, significantly (***) more livestock farmers than mixed, or cropping farmers, in both the UK and Germany agreed with the statement: 'Payments to farmers should be related to regional disadvantage'. This was the statement, it should be remembered, that received the highest level of agreement across the three countries.

Conclusions

It has been shown above that the material from the large scale postal survey of farmers, backed up by a series of on-farm interviews, went against most of the misgivings expressed in the consultations with stakeholders. This is because it was found that there is likely to be relatively little change from the status quo if our proposed changed policy scenario is introduced. The finding from a review of the use of intentions surveys in farming that, if undertaken rigorously, such surveys can accurately portray how farmers will behave, was endorsed by the results of the on-farm interviews which found, generally, that the responses are likely to be reliable. However, whilst the on-farm interviews largely confirmed the intentions of the surveyed farmers, and in that sense validated the survey, we still do not know what these farmers will actually do. Thus, it would be interesting to re-question our respondents in five or ten years' time with the Fischler Reforms well under way, to examine and contrast their actual response with their stated prior intentions.

It has also been shown above that one of the most striking features of our survey results has been how *very similar* the responses of the farmers were. This is regardless of whether they were from the UK, Germany or Portugal, three countries with very different types of farming and rural sectors.

Around 30% of our respondents in each of the three study countries said they would alter their mix of farm activities when support payments were decoupled from current land use and production; the interview phase of the project found that when the full bond scheme was implemented this proportion might rise somewhat. With the exception of forestry and woodland, the spread of proposed new enterprises the farmers said they would undertake on the introduction of the proposed bond scheme was very similar; this finding was markedly skewed by results from Portugal, where not only did proportionately more of their farmers opt for taking up a new activity, but forestry and woodland was the most popular choice.

There were, though, two issues of marked difference between the three study countries highlighted above – that of the likelihood of land abandonment once decoupling took place and the possibility of cashing in of the bond. First, concerning land abandonment, whilst at least 50% in each country said they would leave no land idle, and considerable numbers would idle less than half their land, in Portugal nearly 20% would idle all or more than half their land, a much larger proportion than in Germany and, in turn, this was much more than in the UK. Second, when considering whether farmers would cash in the bond by selling it on the financial market, whilst fairly few across all three countries said they would do this, the German farmers were particularly against doing so in comparison with those in the UK and Portugal. Similarly, the German farmers were much more unlikely to diversify their business when they received their bond than their counterparts in Portugal and the UK.

All in all then, the responses of farmers to the proposed bond scheme, whilst at times somewhat difficult to interpret, have provided, we hope, valuable evidence to suggest that there would be relatively few practical problems at the farm level if it were implemented. Thus, policy makers should not fear the adoption of a bond scheme approach as part of any CAP reform package.

Notes

[1] Copies of the questionnaire used are available from the first-named author.

[2] In order to test whether differences in replies to questions between the various sub-groups were likely to have occurred by chance or otherwise, the responses were tested for statistical significance. The replies were in three distinct forms. First, some replies, or variables, were in a continuous form, such as size of farm or age of farmer. These needed their differences in mean values to be compared using the Students *t* test. Second, some replies were in the form of ordinal scales where, for example, agreement levels with statements were given scores. These were compared using the Mann-Whitney U test. Third, some responses were in a 'discrete' or categorical form (i.e. they had a livestock farm or a crop one). Here, differences in the proportional distribution of replies between such sub-groups were compared using the Chi-square test. If there was no greater chance than 5% that such a large value of *t* or Chi-square or U statistic could have occurred by chance, the difference was stated to be statistically significant. The level of statistical significance will be shown above in the text or in tables as follows:

where there is less than 5% probability that the observed difference would have occurred by chance the mark * will be recorded; where the probability is less than 1% the mark ** will be recorded; and *** where the probability is less than 0.1%. Similarly, where there is a greater than 5% probability that the difference could have occurred by chance, 'ns' will be recorded or that it is 'not significant'.

References

ADAS (2002) *Better Delivery of Government Advice to Farmers.* Land Management Improvement Division, Defra, London.

Bauer, S. (1989) Historical review, experiences and perspectives in sector modelling. In: Bauer, S. and Henrichsmeyer, W. (eds) *Agricultural Sector Modelling.* Vauk Kiel, Germany.

Fearne, A. (1990) Changing practices on the farm - a survey of farmers' perceptions. *Farm Management* 7, 192-198.

Gasson, R. (1983) Part-time farming: a strategy for family farms? In: Tranter, R.B. (ed) *Strategies for Family-worked Farms in the UK.* CAS Paper 15. Centre for Agricultural Strategy, The University of Reading, Reading, UK.

Gasson, R., Errington, A.J. and Tranter, R.B. (1998) *Carry on Farming. A Study of How English Farmers Have Adapted to the Changing Pressures on Farming.* Wye College Press, University of London.

Harvey, D.R. (1990) Agricultural sector modelling for policy development. In: Jones, J.G.W. and Street, P.R. (eds) *Systems Theory Applied to Agriculture and the Food Chain.* Elsevier Applied Science, London and New York.

Harvey, D.R. (2000) *Farmers' Intentions Survey, 1994-1997: Final Report.* Special Studies in Agricultural Economics: No. 47. Department of Agricultural Economics and Food Marketing, The University of Newcastle upon Tyne, Newcastle upon Tyne, UK.

Holt, G.C., Cocker, A.M., Fromant, A. and Martin, A.K. (2001) Changes to the structure of farmers' businesses 1999/2000. In: Vaughan, R.L. (ed) *Farm Business Data 2001.* Department of Agricultural and Food Economics, The University of Reading, Reading, UK.

Johnson, S.R. and Rausser, G.C. (1977) Systems analysis and simulation: a survey of applications in agricultural and resource economics. In: Judge, G.J., Day, R.H., Johnson, S.R., Rausser, G.C. and Martin, L.R. (eds) *Qualitative Methods in Agricultural Economics, 1940's to 1970's,* Volume 2, *A survey of Agricultural Economics Literature.* University of Minnesota Press, Minneapolis, Minnesota.

Johnson, G.L. and Haver, C.B. (1960) *Agricultural Information as an Aspect of Decision Making.* Technical Bulletin 273. Department of Agricultural Economics, Michigan State University, East Lansing, Michigan.

Jones, G.E., Rolls, M.J. and Tranter, R.B. (1987) *Information Management in Agriculture.* British Library R and D Report 5931. Agricultural Extension and Rural Development Centre, The University of Reading, Reading, UK.

Munro, R.G. and Fisher, B.S. (1982) *The Formulation of Price Expectations - an Empirical Study of Wool Growers in New South Wales.* Research Report No 8. Department of Agricultural Economics, University of Sydney, Sydney.

Nielson, J. (1962) *The Michigan Township Extension Experiment: the Farm Families ...their Attitudes, Goals and Goal Achievement.* Technical Bulletin 287. Department of Agricultural Economics, Michigan State University, East Lansing, Michigan.

Poole, A.H. (1983) Dairying on small farms: problems, prospects and practices. In: Tranter, R.B. (ed) *Strategies for Family-worked Farms in the UK.* CAS Paper 15. Centre for Agricultural Strategy, The University of Reading, Reading, UK.

Pryde, J.G. (1982) *Survey of New Zealand farmer intentions and opinions, September-November 1981.* Research Report 126. Agricultural Economics Research Unit, Lincoln College, Canterbury, New Zealand.

Pryde, J.G. and McCartin, P.J. (1983) *Survey of New Zealand Farmer Intentions and Opinions, October-December 1982.* Research Report 136. Agricultural Economics Research Unit, Lincoln College, Canterbury, New Zealand.

Reithmuller, P.C. (1978) Some aspects of information use by pastoral zone graziers. *Quarterly Review of Agricultural Economics* 31, 143-150.

Thomson, K.J. and Tansey, A.W. (1982) Intentions surveys in farming. *Journal of Agricultural Economics* 33, 83-88.

Thornton, D.S. (1961) *Decision Making in Farming, With Exploratory Investigations into the Decision Making of Pig Producers.* Unpublished PhD Thesis, The University of Reading, Reading, UK.

Chapter Nine:

A Role for Direct Payments? The Doha Round, EU Enlargement and Prospects for CAP Reform

Arlindo Cunha
Former Member of the European Parliament, Portugal

Introduction

Direct payments were discussed in Chapter 2, not only in terms of their different nature and role but also their history in the CAP, in US farm policy and in the context of the WTO negotiations. Three main points can be drawn from that analysis. The first is that, progressively, the world's richest economies have been incorporating direct payments for farmers into their agricultural support policies. The second is that the regular use of this practice is recently rooted, dating from the early 1990s, and it cannot be dissociated from the international negotiations aimed at integrating agriculture into the logic of trade liberalisation. The third is that, although direct payments are often justified by the need to supplement and stabilise farm incomes, there are important differences in their nature and effects. The debate on the effects of different agricultural support instruments (direct payments included) has been intensified in recent years at the international level, particularly in the WTO and the OECD (OECD, 2001a and 2001b).

In the early days of the Uruguay Round negotiations the intention was to classify the different types of support by analogy with traffic lights: red, yellow/amber and green. The first should be prohibited, the second would be subject to WTO rules and disciplines and the third could be used freely by Member countries. For practical reasons, this classification was simplified to just two during the negotiation process: those support measures that have a significant impact upon trade (amber) and those that have no or just a minimal impact (green). It was not until the Blair House Agreement that the concept of the blue box was used. The purpose was to recognise the significant contribution that the EU's 1992 CAP reform had made to re-launching the negotiations. It did so by giving special - blue box - treatment to the direct payments which had been introduced to compensate for the price cuts. It is in this context that the Uruguay

Round *Agreement on Agriculture* (URAA) distinguished between three types of support, corresponding to the so-called amber, green and blue boxes.

Amber box support is directly linked with production, consequently affecting farmers' decisions to produce and trade, and for that reason is subject to reduction. Green box payments are mostly decoupled from production, having no or at most a minimal impact on a farmer's decision to produce, and consequently they are not subject to expenditure limits. Blue box payments, whilst linked to land or livestock, form part of 'production-limiting programmes'. They too are not 'subject to the commitment to reduce domestic support'. However, while green box payments are exempted in a permanent way, in that properly crafted green box payments would not infringe other provisions of the WTO agreements, blue box payments would only benefit from a temporary derogation corresponding to the nine years' duration of the Peace Clause (Article 13 of the URAA), and then only to the extent that 'such measures do not grant support to a specific commodity in excess of that decided during the 1992 marketing year'.

The evidence from recent experience in the implementation of agricultural policies is that the more developed countries started to adopt direct payments for farmers on a regular basis when their room for manoeuvre to use market and price policies to support farmers' incomes came under attack in the multilateral negotiations to liberalise agricultural trade. It can be concluded, therefore, that negotiation in the WTO has exerted strong pressure upon the configuration of agricultural policy reforms. In the EU, direct payments were able to reconcile the purpose of securing farm incomes at a reasonable level with restrictions on the use of certain farm policy instruments in the GATT/WTO, and for the need for simplification given the prospect of EU enlargement. Thus they are fated to have a very important role in the future, not only in the EU but as part of the agricultural policies of many richer countries.

But what form should these direct payments take? According to OECD and WTO criteria they should be decoupled from production. However, such a principle is far from being unanimously accepted by some key WTO partners, notably the EU and its allies. The main objective of this chapter then is to review the debate on direct payments in the context of the process of CAP reform and its links with EU enlargement and the WTO negotiations.

The process of CAP reform

Proposals in the mid-term review

The possibility of introducing new changes into the CAP before the end of the implementation period (2000-2006) of the existing policy was envisaged at the time of the Agenda 2000 reform in March 1999. The Commission's 'first task' in the mid-term review was for a 'stock-taking' of the Agenda 2000 reforms, and to respond to the European Council's invitation to suggest improvements to the policy, 'where appropriate, in order to ensure that the objectives of the Agenda 2000 reform could be fully realised'. In addition, at the Göteborg European Council, the Commission had been asked to consider 'the effects on environment and sustainable development'. Thus the Commission was able to claim that 'the mid-term review ... provides a unique opportunity to achieve the objectives set in Berlin and Göteborg, and to respond to the high expectations of European citizens with respect to agriculture and agricultural policy' (Commission, 2002)[1].

From its review the Commission concluded that further changes to the CAP were warranted, and - with particular relevance to the theme of this book - it suggested, *inter alia*, that:

1. Agricultural production must be more orientated to the products and services that the public wants and not to artificially created price incentives or product-specific aids. Direct income payments should not steer the production decisions of farmers.
2. Support and stabilisation of agricultural incomes remains an essential objective. Direct payments must therefore continue to play their role in promoting a fair standard of living for the agricultural community.
3. Support between the two pillars of the CAP must achieve a better balance to meet society's expectations of a policy that promotes food quality, sustainability, and value for money through reinforced rural development programmes (Commission, 2002).

The Commission's July 2002 Communication contained sector-specific reform proposals for cereals, rice and nuts, four possible options for the milk regime, and the introduction of a new 'carbon credit' payment of €45 per hectare for producers of energy crops.

The real innovation consisted of two main measures. First, a further decoupling of support with the introduction of a single decoupled payment (now known as the Single Payment Scheme) to replace the existing arable area and headage payments. The decoupled payments would be based on historic entitlements, but payment would be conditional on compliance with statutory rules on environmental, food safety, animal health and welfare, and occupational safety (a linkage known as 'cross compliance'). Having coped with such principles, farmers would then be free to produce what they found to be more profitable as they responded to market demands, or to produce nothing provided they ensured that land was maintained in good agronomic condition. In order to facilitate the mobility of land and farmers, it was proposed that the single farm payment would be divided by the eligible area of each farm and give rise to a certain number of payment entitlements (expressed in hectares), which could be sold with or without the land.

The single payment would embrace existing payments for arable crops and livestock, and the dairy premium from 2005 agreed as part of the Agenda 2000 reforms. In some sectors, however, the decoupling would not be complete, in that a premium coupled to production was proposed as a supplement to the basic decoupled aid incorporated in the Single Payment Scheme. That was the case for durum wheat (a quality supplement), rice, protein crops, flax, hemp, dried fodder and starch potato.

The second innovation was *dynamic modulation* under which all direct payments would be progressively reduced, starting with a 3% reduction and rising by 3% a year up to a maximum of 20%. The total amount saved would be redistributed to Member States on the basis of agricultural area, farm labour, the level of income (cohesion criteria) and would reinforce the CAP's Second Pillar. For each recipient, the first €5,000 would be exempt from modulation. This franchise could be increased by €3,000 for each additional AWU above two at the request of a Member State. The Commission (2002) noted that the franchise 'will ensure that the majority of farms will not be subject to modulation'. Last, but not least, a cap of €300,000 per applicant was proposed, with funds also transferred to the Second Pillar. However, in contrast with the budget savings generated by modulation, which were to be redistributed on Pillar 2 activities across the EU, the funds made available from capping would be retained for use within the Member State concerned.

Besides these two major proposed changes - which are emblematic of the Commission's MTR proposals - there were other *horizontal* measures. Mention has already been made of the proposals on cross-compliance. In addition, *environmental set-aside* would roll-forward the existing requirement to set-aside 10% of arable land under the arable area payment scheme, but would do so on a non-rotational basis. A new farm auditing system would be funded, aimed at helping farmers cope with the new requirements and standards. Finally, reflecting its increased funding, the concept of the Second Pillar of the CAP would be considerably enhanced, with the integration of animal welfare, food safety and food quality preoccupations within its scope.

As one might expect, given the radical nature of the proposals, the reaction of the overwhelming majority of the Member States was cautious and hostile. Many ministers claimed that what the Commission was proposing was not a *review* as agreed in Berlin, but a *radical* reform (see, for instance, Agra Europe, 2002). We return below to the specific concerns raised by the proposals.

A long-term policy perspective for sustainable agriculture

Five months later, and after having received reactions from Member States, EU institutions and interest groups, the Commission presented its legislative proposals, now under the heading *A Long-term Policy Perspective for Sustainable Agriculture* (Commission, 2003a).

For the most part the January 2003 proposals mirrored the July 2002 text, although a new element was a specific proposal for reform of the milk regime. In particular, it retained the proposal for a single decoupled payment, with only a minor adjustment to the treatment of potato producers.

However there were substantial changes to dynamic modulation. The term was dropped, to be replaced by *degression*. This would involve the introduction of a progressive and differentiated rate of modulation according to the level of aids received by farmers, rising from 1% in 2006 to 19% in 2012, but with the preservation of the €5,000 franchise. The proposed €300,000 cap on payments was dropped. Furthermore, only about one-third of the funds reallocated through degression would be allocated to rural development, with the remainder used to finance new CAP reforms.

The final compromise on CAP reform was reached on the morning of 26 June 2003 at the Agriculture Council in Luxembourg, after a marathon session which had started on 11 June and had twice been suspended (Agra Europe, 2003b). With regard to the elements being tracked in this chapter, the final compromise had the following components.

The concept of a single decoupled payment was accepted for the sectors proposed by the Commission, from 2005. However, either for reasons related to the specificity of some sectors, or the fear of some Member States about the abandonment of production in less competitive regions, this principle is tempered with three variants of *partial decoupling*. The first is the payment of production-specific supplements for some crops: durum wheat, protein crops, rice, fodder, flax hemp and potato starch. The second is the exemption from decoupling of area payments on seeds, dried fodder and of all the payments in the outermost regions of France (the overseas territories), Portugal (Azores and Madeira) and Spain (Canaries). The third allows Member States to preserve coupled payments up to a certain level in the sectors otherwise subject to full decoupling. These can be 25% in the arable sector (or 40% for the production-specific supplement to durum wheat), 50% for sheep and goats and 60% for starch potatoes. For the beef sector Member States have three options: retain the link with production for 100% of the suckler cow premium and 40% of the slaughter premium; *or* 100% of the slaughter premium and nothing else; *or* 75% of the male beef special premium. Member States are only obliged to implement the decoupled system after 2007, and they can deduct up to 10% of the total amount of the payments and use the funds to encourage farming activities that are important in protecting and enhancing the environment, and to improve the quality and marketing of agricultural products.

Modulation was fixed at 3% in 2005, 4% in 2006 and 5% for the period from 2007 to 2013 for annual payments in excess of €5,000. However, at least 80% (90% in Germany) of the modulated funds will be retained in the Member State for Pillar 2 activities, and very little will be redistributed between the Member States according to the criteria proposed by the Commission. From 2007, if budget funds are tight, and so as to keep within budget ceilings agreed in October 2002, the Commission has been authorised to reduce the level of direct payments, but subject to modification by the Council.

The main change on cross compliance was a substantial simplification of the proposed scheme, with a reduction from 40 to 18 regulations that are to be integrated in the compulsory scheme, a softening of the penalty conditions, and the retention within the Member State of 25% of the fines imposed.

The dynamics of the negotiations

There were two main players in the political process of this CAP reform: the Commission, led by Commissioner Franz Fischler, and the France-Germany axis.

The obstinacy of the Commissioner stemmed from his dissatisfaction with the final outcome of the Agenda 2000 reform, in March 1999, which fell short of his initial proposals in its three respects: price cuts; rural development; and the financing of the CAP. The price reductions agreed were lower than those proposed by the Commission, and for milk only applied from 2005. To reinforce the CAP's Second Pillar, and abate the agricultural budget problems, the Commission had proposed the introduction of modulation/degressivity and a capping of direct payments. After lengthy discussion in both the Agriculture Council, and at the European Council in Berlin, the package that emerged differed significantly from the Commission's proposal, whilst the Commission's emblematic idea of creating the Second Pillar was rescued from failure at the last minute by postponing milk reform thereby reducing budgetary constraints.

The idea of a mid-term review for the CAP must, consequently, be seen in the context of the Commission's evaluation of the Agenda 2000 reform. This was deemed to be insufficient to match internal and external challenges, in that in essence it retained the status quo policy model.

During 2001 the Commissioner and his top-ranking officials fuelled discussion on a new reform. This discussion was, however, balanced between the extremes of a simple review and a true reform. The justification given for the need for a new substantial reform was based on both internal and external factors. The basic internal preoccupation was to maintain market balance for the more important commodities, namely beef (with intervention stocks piling up as a result of the BSE crisis) and cereals. The external pressure was derived from the WTO negotiations for a new round. Addressing the European Parliament just after the WTO Ministerial Conference at Doha, Franz Fischler bluntly noted that 'Doha isn't dictating reform (...) but it has to be taken into account. We can't keep the same level of export refunds forever' (Agra Europe, 2001b).

It should be said, however, that the Commissioner's approach to the MTR continued to be essentially that of policy adjustment, weighed with his usual preoccupations about rural development, quality and international commitments. This is clear in his address to the Confederation of European Agriculture's Conference in Belfast in September 2001: 'As you can see, the objectives of Agenda 2000 are still valid. This means that our mid-term review, which we have to carry out next year is not going to be another reform. A reform would be if we wanted to change our objectives and the financial framework. But what we have to do is to review the tool kit that we have implemented with Agenda 2000 and to see whether it is fit to reach our objectives' (Fischler, 2001). In contrast with these ideas, Dirk Ahner, Deputy Director-General of DG Agriculture and considered one of the main *brain-aids* to Commissioner Fischler, at the Agra Europe Outlook Conference earlier in the year had commented that the forthcoming MTR would provide 'a unique opportunity for stocktaking and a new debate on the future reorientation of the CAP' (Agra Europe, 2001a). Such a broader perspective undoubtedly contrasts with the Commissioner's light view of the reform.

To evaluate to what extent these two positions are really contradictory is beyond the scope of this chapter. Two ideas, however, seem plausible. The first is that the Commissioner's words in Belfast reflect the fact that he was speaking to farmers' leaders where a popular and relaxed intervention would be expected. The second is that the Commission never lost sight of proposing a bold reform in the context of the MTR, but in political terms the Commissioner could not take risks by prematurely disclosing his hand, because an extemporaneous discussion could *kill* the project. The idea then was to work on the project within a limited group of the Commissioner's trusted officials, and evaluate at the same time the political reactions from the Member States.

There are two objective reasons to explain the progressive evolution of the Commission's view of the MTR, from mere adjustments to market policy to a much broader policy reform. The first was the simplification challenge imposed by enlargement. The second is related to the calendar and contents of the WTO negotiations which soon revealed the tough positions of the US and the Cairns Group, with their insistence not only on the elimination of export subsidies (which would imply, in the limit, substantial cuts in the internal price) but especially the end of the blue box and its transformation into the amber box. The Commission realised that being forced to defend the blue box would put the EU in a defensive position, thus jeopardising its ambitions in the negotiation process.

The year 2002 had two main challenges. The first was that rumours from Brussels that DG Agriculture was working on a radical reform plan rang alarm bells in France, where President Jacques Chirac faced re-election. Fischler went to Paris and reassured the French President that no reform proposal would be presented before the presidential election in May and the parliamentary election in June. The deal was respected, but the Commissioner did not give up his broad interpretation of the Commission's mandate for the mid-term review.

When discussions started, the Commission was able to organise an effective communications campaign explaining the need for radical reform. But with the re-election of Jacques Chirac in May, and the return of a centre-right government in France, things became more difficult for the Commission. France was opposed to decoupling, and indeed to any idea of changing the level or method of support, before the end of the Agenda 2000 implementation period (2006).

France and Germany once again played a key role, independent of domestic party politics, in setting the framework for the negotiations. This was evident in four phases. First, as early as summer 2001, the two farm ministers (the green Renate Künast and the socialist Jean Glavany) published an article in *Le Monde* in support of a *reorientation* of the CAP which would reinforce the funding of rural development by making compulsory the modulation of direct payments (Glavany and Künast, 2001). In a second phase, the French-German axis worked at the highest political level to settle the framework for the future financing of the CAP in the Brussels European Council of October 2002 (discussed below). In a third phase, the two delegations worked hard to reach an agreement on the proposals. Despite the huge difference of views, this was possible on the eve of the Council meeting of 11 June. Lastly, the French-German axis forced a suspension of the Council session on 20 June 2003.

The end result was that both countries largely achieved their objectives. Germany had a real commitment to reform the CAP and progressively reduce the agricultural budget, but could not achieve this with the opposition of France. France wanted to minimise the impact of a reform which it knew was unavoidable. However, for that to be possible it needed the understanding of Germany, especially because it did not have the support of the Commission.

The main debates

Aside from purely political discussions dealing with such things as the timing of the reform and links with the WTO negotiations, the proposals for milk and cereals, and other national preoccupations, there were two main strategic debates which cut across all other issues: one on the proposed system of decoupled payments; the other on budgetary concerns, which in turn was closely associated with the debate on modulation, degressivity and rural development.

The debate on decoupling was probably the most passionate, because it provided the main ground of opposition to the reform from almost all the farmers' organisations and from many Member States, and because it was around this that the compromise package had to be agreed. This was the core of the Commission's proposal.

The advantages of decoupling had been strongly advocated by the Commission: farmers would receive a financial cushion to help them support the high production costs they face as a result of society's expectations of their stewardship of land, given the increased competition - lower market prices - brought about by WTO agreements; and at the same time they would be free to produce what they found most profitable given market prospects and the productive capacity of their farms. Besides its economic rationality, the new system would entail an enormous simplification in the implementation of the CAP: a simplification which would become more urgent for an enlarged EU.

Against those arguments, farmers and some Member States raised three basic criticisms. The first was the risk of farmers giving up production given that the complete separation between production and payments would imply that farmers could receive payments even if they took the decision to produce nothing. This risk was particularly serious in marginal areas[2]. Farmers feared a negative reaction of European public opinion to the idea of *paying farmers for doing nothing*. This could create future problems for continued public support for the CAP, even in a reformed shape. Another fear was that it would put in danger the multifunctional system of European agriculture. As in most cases where there is *joint production* of *market* and *public* goods, the abandonment of production (the *market* function) in certain areas would imply also the end of the other (*non-market*) functions and consequently the economic and social *death* of those regions (this concept and an analysis of its applicability to the EU is expressed in Massot (2001)).

The second main criticism related to the adoption of an historic base as the criterion for the calculation of future levels of decoupled payments for farmers. This meant:

- the perpetuation of the existing distributive inequities of the CAP; and thus
- the continued discrimination between products, farmers and regions. The level of CAP support to each farmer would forever approximate what they received in the past.

This would imply the perpetuation of discrimination against those farmers producing products not entitled until now to receive direct aids, including most of the Mediterranean products. They have not enjoyed a generous level of price support in the past. For that reason they were not included in the 1992 and 1999 CAP reforms and thus were not entitled to receive compensatory aids. Consequently they would be now excluded forever from receiving a direct aid, despite facing the same production costs as their fellows who are receiving direct aids and fulfilling the same multifunctional role towards society.

The situation just described was well pictured in Fig. 5.2, on page 84. Two conclusions seem evident: that the countries with lower levels of income per labour unit (in general the Mediterranean countries) are those enjoying a lower level of CAP support; and, consequently, that the CAP does not work as a cohesion instrument as recommended by Article B of the Amsterdam Treaty (now Article 2 of the Treaty of the European Union), but rather the opposite[3].

The third criticism of decoupled payments related to the distortion of competition between products. In the Commission's Communication of July 2002, farmers' freedom to produce was unlimited. Let us imagine two neighbouring farmers. One had always produced wine or fruit and vegetables, and consequently would not be entitled to a decoupled single payment. The other used to grow oilseeds or cereals, and would be entitled to receive a decoupled payment into the future. However, as the aid is decoupled, the second farmer could decide to give up cereal or oilseed production and instead plant a vineyard, or an orchard, or grow fresh vegetables. The result will be that both farms produce the same products; but with a difference. One receives direct subsidies and the other does not. It is then not difficult to conclude that such a distortion of competition is unfair and, as such, untenable. Recognising the pertinence of the argument, the Commission changed its proposals, first to exclude permanent crops in the January 2003 text, and then fresh vegetables, including potatoes, in the final compromise. The final outcome also allows Member States to implement the scheme on a flat rate, regionalised, basis which will facilitate payment over a larger area.

The budgetary implications of the reform were also a cause for debate. The Commission's July 2002 Communication had a basic weakness: the proposed reform entailed a budget problem from 2007 onwards. The solution was found in the decisions of the European Council of October 2002 held in Brussels. This set a new financial guideline at the level of 2006 expenditure (€45,306 million) plus 1% each year to reach €48,574 million in 2013, for EU-15. Even though the European Council made it clear that expenditure for the Second Pillar (category 1-b of the budget) was not included in the new guideline, the decision created a serious restriction on Fischler's ambitions for rural development. Although the decision was taken by unanimity, it was led by Germany and France who had agreed the deal two days before. France, which had never been an enthusiast for rural development, succeeded in postponing the implementation of modulation until 2005 and Germany secured the guarantee that agricultural expenditure would decrease in real terms in the future. The Commission had then to incorporate these decisions in its formal legislative proposals for the MTR, with two consequences. First, its proposal for *degression* to finance the reform; and second the reduced transfer of funds from Pillar 1 to Pillar 2 to less than one-third of the Commission's original ambition.

Two important conclusions stem from comparing the contents of the Commission's July 2002 and January 2003 presentations of the MTR: the radical change in the contents of the proposals relating to modulation; and the dogged insistence to retain its proposal for decoupled payments. The explanations for such behaviour are certainly complex and difficult to interpret. However, it seems legitimate to conclude that the implementation of a new system of direct payments was the leitmotiv of this reform; not for the sake of the payments themselves, but for their impact on the enlargement process and the WTO negotiations.

The context and pressures for CAP reform

The mostly frequently raised question about the motivation for the 2003 CAP reform is: *Why did the Commission present a real (and substantial) reform proposal when, by the political commitments established by the Council, it was only obliged to work on a 'review'?* The Commission's response to such a question was twofold. On the one hand it argued that the Council's mandate did not limit the Commission's capacity of initiative. Second, it claimed that this initiative was justified because of the need to prepare the European farm sector and its agricultural policy for new and fundamental challenges, such as:

- to bring farmers' production decisions closer to consumer demand;
- to increase the policy's capacity to provide *public services* (reinforcement of the Second Pillar), thus legitimising the CAP to society;
- to facilitate the multilateral trade negotiations in the WTO; and
- to better address the challenges of enlargement.

The influence of enlargement

There are two basic reasons to suggest why CAP reform was seen as necessary to facilitate enlargement. The first was the need to simplify the system of direct payments in which the number of farmers (EU-27, from 2007) would more than double. Not only would the number of farmers raise administrative problems for implementation and control, but there is also the fact that in the Accession and Candidate States there is a general lack of basic information relating to property identification. It was in view of such difficulties that the Commission recommended that direct payments to farmers in the new Member States could be decoupled from production and implemented in a flat-rate simplified way. Although not mandatory, it is very important to realise that Poland quickly decided to adopt this scheme and that most of the other new Member States are likely to do the same.

The second reason relates to budgetary problems. It was widely accepted that the EU could not simply extrapolate the existing financial effort to support its agriculture in an enlarged EU. It was this need to stabilise farm expenditure in an enlarged EU that was at the heart of the difficulties to negotiate the phasing-in of direct payments in the new Member States. In the medium-run the problem has been solved with the phasing-in of direct payments, as outlined in Chapter 2, but new solutions are necessary after 2012.

As a result the only way to continue with the CAP reform process in an enlarged EU (involving further price cuts, and new commodity regimes, *and* compensation in the form of direct payments) is to cut the existing direct payments received by farmers. It follows then that enlargement (and the budgetary pressures associated with it) worked as a shaping element of this reform, besides explaining its poor contents concerning the reinforcement of rural development. Table 9.1 clearly displays that without modulation a financing problem for the CAP would have reappeared by 2009.

The WTO dimension

One feature of the URAA was the recognition in Article 20 that *'the long term objective of substantial progressive reductions in support and protection resulting in fundamental reform is an ongoing process…'* It goes on to state that one year before the end of the implementation period WTO Members were due to start negotiations *'for continuing the process'* taking into account *'what further commitments are necessary to achieve the above mentioned long-term objective'*.

Table 9.1. Financial framework for heading 1a for the period 2007-2013, and the implications of the Commission's January 2003 MTR proposals
(Million euros, current prices)

	Base 2006	2007	2008	2009	2010	2011	2012	2013
I. Budget limit set October 2002								
	45,306	45,759	46,217	46,679	47,146	47,617	48,093	48,574
II. Estimated spend on unreformed CAP for EU-25								
	44,749	45,659	46,734	47,467	48,257	48,852	49,448	50,044
III. Margin (I-II)								
	557	100	-517	-788	-1,111	-1,235	-1,355	-1,470
IV. Estimated spend with EC's January 2003 reforms, but without degression								
	44,395	45,156	46,123	47,568	48,159	48,805	49,451	50,099
V. Revised Margin (I-IV)								
	911	603	94	-889	-1,013	-1,188	-1,358	-1,525
VI. Net proceeds from degression								
		228	751	2,030	2,420	2,810	3,200	3,343
VII. of which available for Rural Development								
		228	475	741	988	1,234	1,481	1,481
VIII. Revised Margin for 1a Expenditure (V+VI-VII)								
	911	603	370	400	419	388	361	337

Source: Adapted from European Parliament (2003)

It should be said that Article 20 also states that the negotiations should take into account other factors including past experience and the effects of implementation of the *Agreement on Agriculture*, non-trade concerns, and special and differential treatment for developing countries. It was in this context that the WTO Ministerial Conference at Seattle, in 1999, attempted to set an agenda and a time schedule for the negotiations. Despite all the efforts made, the different parties were much divided by their interpretations of Article 20, and for this and other reasons the final result of the Seattle Ministerial was an extensive failure.

However, at the Doha Ministerial Conference in November 2001, agreement was reached to embark on a new round of agricultural trade negotiations, in the context of the Doha Development Agenda. The relevant text from the Ministerial Declaration essentially re-states the contents of Article 20, as follows:

> 'We recognise the work already undertaken in the negotiations initiated in early 2000 under Article 20 of the Agreement on Agriculture, including the large number of negotiating proposals submitted on behalf of a total of 121 Members. We recall the long-term objective referred to in the Agreement to establish a fair and market-oriented trading system through a programme of fundamental reform encompassing strengthened rules and specific commitments on support and protection in order to

correct and prevent restrictions and distortions in world agricultural markets. We reconfirm our commitment to this programme. Building on the work carried out to date (and without prejudging the outcome of the negotiations)[4] we commit ourselves to comprehensive negotiations aimed at: substantial improvements in market access; reductions of, with a view to phasing out, all forms of export subsidies; and substantial reductions in trade-distorting domestic support. We agree that special and differential treatment for developing countries shall be an integral part of all elements of the negotiations and shall be embodied in the Schedules of concessions and commitments and as appropriate in the rules and disciplines to be negotiated, so as to be operationally effective and to enable developing countries to effectively take account of their development needs, including food security and rural development. We take note of the non-trade concerns reflected in the negotiating proposals submitted by Members and confirm that non-trade concerns will be taken into account in the negotiations as provided for in the Agreement on Agriculture.

Modalities for the further commitments, including provisions for special and differential treatment, shall be established no later than 31 March 2003. Participants shall submit their comprehensive draft schedules based on these modalities no later than the date of the fifth session of the Ministerial Conference. The negotiations, including with respect to rules and disciplines and related legal texts, shall be concluded as part and at the date of conclusion of the negotiating agenda as a whole' (WTO, 2001: paragraphs 13 and 14).

There are three real innovations. The first, the reference to the *phasing out* of export subsidies as a possible goal of the negotiations, was imposed by the WTO Director General with the support of the US and the Cairns Group. The second, the qualification that all the goals referred to in the declaration are stated *without prejudging the outcome of the negotiation*, was imposed by the EU and its *friends of multifunctionality,* a group of strategic allies of the EU in the WTO negotiations formed by the Central and Eastern European candidates for EU membership, Norway, Switzerland, Japan and South Korea. The third new element is the definition of a time schedule for the remaining stages of the negotiations. The 'modalities', that is to say the new rules of the successor agreement, were to be agreed by 31 March 2003; Members were to table their draft schedules before the Fifth Ministerial Meeting to be held in Cancún, Mexico, in September 2003; and - not referred to in the text cited above - the whole of the Doha Agenda was to be concluded as a single undertaking before 1 January 2005.

During the course of 2002, and into 2003, there were few surprises as WTO Members outlined their thoughts on the modalities. The EU suggested a continuation of the basic rules of the URAA, with some commitments at an even higher level (an average tariff reduction of 36%, a 45% cut on *all types* of export subsidies, and a 55% reduction in the aggregate measurement of support (AMS)), a maintenance of the status quo in relation to the contents of the three boxes, an end to the *de minimis* clause and more flexible provisions to deal with non-trade concerns (Commission, 2003b).

The US proposed a new formula for tariff cuts (the so-called *Swiss-25 formula*, which would reduce the maximum tariff rate to 25%), elimination of the special safeguard clause, elimination of export subsidies over five years, and limiting domestic support (both blue and amber boxes taken together) to 5% of the value of agricultural production. The Cairns Group followed a line close to that of the US, with one substantial difference. They suggested the elimination of the *de minimis* clause that the US - which has made extensive use of this provision - wished to see retained. It should be stressed that, in 1999, the US spent about $8,000 million under this clause (exempted from reduction commitments), while having notified only $16,000 million under the amber box reduction commitments (intervention of Commissioner Franz Fischler in the Agriculture Council of 17-18 February 2003).

The deadline of 31 March 2003 for securing agreement on the modalities was not respected, partly as a result of the late presentation of the offers of some Members (including the EU), and partly because of the late tabling of the draft document by the Chairman of the WTO Agricultural Committee, Stuart Harbinson. A revised draft was made available just a few days before the special session of 24 March 2003 (WTO, 2003; see also Ruffer and Swinbank, 2003). Making it clear that the document was submitted on his own responsibility with the aim of bridging differences of views among the Members, Harbinson's text focused on the three main topics of the agricultural negotiations.

On *domestic support* the document preserved the rules governing the amber and the green boxes, and proposed a 60% cut for the former and more restrictions on emergency payments under the latter. Much different is the fate of the blue box, for which two options were presented: a 50% reduction over five years; or its transformation into the amber box. The *de minimis* clause would be subject to an annual reduction of 0.5 percentage points over the five year implementation period (i.e. it would be halved, from 5 to 2.5%), whilst retaining it at 10% for developing countries. In addition, the document also proposed an exemption for developing countries from reduction commitments of certain amber box type support instruments, such as payments related to basic crops if they are important for food security purposes, or 'Payments to small-scale producers/family farms for the purpose of maintaining rural viability and cultural heritage in developing countries' (WTO, 2003: Attachment 9).

As might be expected from a document of such a complex nature, the reactions of the different parties could not be other than tactical in order to enlarge their room for manoeuvre for the negotiations phase. One of the more critical reactions came from the EU's Franz Fischler who considered the paper disappointing. It was unbalanced and incapable of bridging the differences between WTO Members. His reaction is captured in the following declaration:

> 'In the developed world, those that moved in a direction that is consistent with what was agreed in the previous round are penalised, while those that reversed direction get rewarded. In the developing world, those that need a boost to reap the benefits from trade find themselves treated equally with those more developed, or fully developed. And those concerned about enhancing the provision of public goods, from environment to food safety, see their non-trade concerns ignored and their concerns about the potentially negative impact of trade enhanced' (Fischler, 2003).

Although not directly expressed, there is no doubt that these comments were addressed to the US following its approval of the Farm Security and Rural Investment Act (FSRIA) which the EU saw as a clear setback vis-à-vis the liberal and decoupled contents of the 1996 FAIR Act (Haniotis, 2002).

In comparing the EU's position paper with the Harbinson draft, the only point on which the two seem close is on the linear method for tariff cuts, clearly opting against a formula approach for the elimination of tariff peaks. However, on most of the remaining key issues the Harbinson draft seems to be closer to the views expressed either by the Cairns Group or by the US: partial or total elimination of the blue box; insistence on a timetable to phase out export subsidies; elimination of the special safeguard clause, but retention of a *de minimis* clause; lax treatment of export credits; insufficient distinction between the special and differential treatment to be given to Least Developed Countries and to the more general class of developing countries, and no consideration of the special treatment to be afforded the ACP (African, Caribbean and Pacific) States that have had an association status with the EU since the 1970s (now expressed in the Cotonou Accord); and a disregard of geographical indications of origin and the generality of non-trade concerns.

Even though all of the above are sensitive negotiating issues, market access, export subsidies and domestic support continue to be the key and will undoubtedly be the most difficult ones to agree. It is not difficult to find arguments from other WTO partners with almost opposite views, especially bearing in mind that we are in a negotiation characterised by a bargaining game between the different interests. Speculation on whether an equilibrium can be reached in Cancún, or even before January 2005, is a matter for futurology.

Prospects for direct payments in future negotiations

The context described above makes clear that direct payments are again (as they were in the URAA) at the epicentre of the ongoing WTO negotiations, and are subject to a two level game: in the WTO and in the EU.

Although there is some room for dispute about the size of the amber box and the content of the green box, these are not particularly contentious issues. Completely different is the discussion about the blue box. Three points could be made in this respect. The first is that, even though the blue box is currently a permanent provision of the URAA, implying that it can only be removed through negotiation, it is only safe from challenge under the Subsidies Code and other WTO provisions for the duration of the Peace Clause. This is what gives the EU's main competitors leverage when they demand that blue box support should be rolled into the amber box.

The second point is that although it is clear that blue box payments are not sufficiently decoupled to meet the WTO's green box requirements, it is also true that they cannot be equated with amber box product-specific payments for the simple fact that they are limited to historic levels and consequently do not have the same distorting effects.

Thirdly, the EU itself has fuelled confusion as a result of the constraints inherent in its decision-making process. The European Commission has in fact presented two apparently contradictory positions as far as blue box payments are concerned. In the WTO, it tabled an offer that would preserve 100% of its blue box; but within the EU it presented a proposal for CAP reform which implied the full elimination of the same blue box. The explanation for such an apparent contradiction is that the Commission could not make a WTO offer that matched its proposals for CAP reform in the MTR because of the limits of its mandate for the WTO negotiations agreed in the Council. However, by making known its reform proposal, all of the EU's negotiating partners in the WTO became aware that the Commission (if not the Council) was willing to get rid of the blue box.

It should also be noted that the EU has considerable room for manoeuvre in the face of its WTO commitments, implying that it did not necessarily need to decouple all direct payments (see Table 9.2).

Table 9.2 reports estimates computed by the author for a European Parliament report. It begins by recording the blue and amber box declarations made to the WTO by the EU for the year 1999/2000. Amber box support, at €47.9 billion, fell short of the maximum allowed (€69.5 billion), leaving a considerable margin (€21.6 billion). Even had blue box expenditure (€19.8 billion) been included in the amber box, the WTO ceiling would not have been breached. It is estimated that the Agenda 2000 reforms, agreed in March 1999, will boost blue box expenditure in EU-15 by €9.6 billion, whilst reducing amber box support by nearly €20 billion, thus increasing the EU's scope to absorb new Member States and agree, in WTO negotiations, to a further reduction in its AMS ceiling. The Commission's January 2003 MTR proposals would have switched most blue box expenditure into the green box, further enlarging its room for manoeuvre in the WTO.

Even though the decisions taken by the Council of Ministers of Agriculture watered down the Commission proposals, the 2003 CAP reform implies that about two-thirds to three-quarters[5] of the blue box is shifted to the green box. Even though this result differs from the radical positions of the USA and the Cairns Group, calling for an elimination of the EU's blue box, it still is a far reaching achievement, for two reasons.

Table 9.2. Estimates of EU-15 support subject to reduction commitments in the WTO (million euros)

	Support notified to the WTO for 1999/2000 G/AG/N/EEC/38	Support after Agenda 2000 (including the dairy reform) Estimate	Support with the Commission's Jan. 2003 MTR proposals Estimate
a) Blue Box	19,792	29,406 *of which:* - arable 17,466 - livestock 11,940	468 (product specific payments)
b) Amber Box (AMS)	47,886	28,590	25,959
c) Ceiling available	69,463	67,159 (for 2000/01)	67,159 (for 2000/01)
d) Margin (c-b)	+21,577	+38,569	+41,200
e) Total Support (a+b)	67,678	57,996	26,427

Rapporteur's estimates. Source: Adapted from European Parliament (2003)

The first is because it gives the EU a comfortable position in the forthcoming negotiations. With decoupled payments now approved, the EU enhances its bargaining capacity over other problematic issues, such as: the flexibility it wants in green box instruments to retain multifunctional agriculture over its entire territory; protection of geographical indications of origin; and other non-trade concerns such as animal welfare. There is also a tactical advantage. The Commission knows that it does not need to eliminate the entire blue box - after all, the Harbinson paper only proposed a 50% reduction. However, the Commission also knows that it is only by making a principled stand on a number of issues that it will be able to get something at the end of the negotiations.

The second reason is that, with this decision, the EU sent a clear message to its farmers and trade partners about its future intentions concerning direct payments. Decoupled direct payments are destined to become a more central feature of the CAP, and one might even envisage additional features of the bond scheme being incorporated in future CAP reforms. On the other hand it will be important to find policy mechanisms that preserve the multifunctional characteristics of European agriculture; and it should perhaps be stressed that the retention of a blue box facility in the WTO could be a very useful instrument to facilitate policy reform in other sectors.

Notes

[1] The Berlin European Council of March 1999 decided the overall package of Agenda 2000, including a wide-ranging CAP reform which continued the 1992 reform and was envisaged as one step more in an ongoing process. It also set a budgetary framework for the period 2000 to 2006. The 2001 Göteborg European Council approved a European strategy for sustainable development according to which the economic, social and environmental effects of the common policies should be fully assessed in a coherent and integrated framework and taken into account in the EU decision-making process.

[2] Impact studies have been carried out by the Ministries of Agriculture of Portugal and Spain. In the case of Portugal it was concluded that 14% of the farms representing 46% of the utilised agricultural area (UAA) and 30% to 40% of the sheep and beef sectors could run that risk (MADRP, 2003). The Spanish study concluded that about 1.75 million hectares of UAA could be abandoned, particularly in the regions of Castilla la Mancha, Aragon and Extremadura (MAPA, 2003).

[3] Article 2 of the Treaty of the European Union reads, *inter alia*, 'to promote economic and social progress and a high level of employment and to achieve balanced and sustainable development, in particular through the creation of an area without internal frontiers, through the strengthening of economic and social cohesion and through the establishment of economic and monetary union, ultimately including a single currency in accordance with the provisions of this Treaty'.

[4] The phrase in brackets was added during the negotiations to the original draft Declaration prepared by the WTO Director General. The brackets are not in the WTO text.

[5] Given that the Member States can implement differently the decoupled system of payments, it is not possible to calculate the exact amount of the reduction of the blue box. Preliminary estimates indicate that the blue box could be reduced to €8-10 billion, implying a reduction of between 66% and 73% (Agra Europe, 2003a and 2003c, and own estimates).

References

Agra Europe (2001a) 30 March, A/1.
Agra Europe (2001b) 23 November, EP/4.
Agra Europe (2002) 19 July, EP/1-EP/8.
Agra Europe (2003a) 26 June, A/2.

Agra Europe (2003b) 27 June, EP/1-EP/11.

Agra Europe (2003c) 4 July, A/1.

Commission of the European Communities (2002) *Communication From the Commission to the Council and the European Parliament. Mid-term Review of the Common Agricultural Policy.* COM(2002)394. CEC, Brussels.

Commission of the European Communities (2003a) *Proposal For a Council Regulation Establishing Common Rules For Direct Support Under the Common Agricultural Policy and Support Schemes For Producers of Certain Crops ... (A Long-term Policy Perspective For Sustainable Agriculture).* COM(2003)23. CEC, Brussels.

Commission of the European Communities (2003b), *WTO and Agriculture: the European Union Takes Steps to Move the Negotiations Forward.* Press Release IP/03/126, 27 January 2003. CEC, Brussels.

European Parliament (2003) Committee on Agriculture and Rural Development. *Report on the Proposal for a Council Regulation on Establishing Common Rules for Direct Support Schemes Under the Common Agricultural Policy and Support Schemes for Producers of Certain Crops* (Rapporteur Arlindo Cunha). Document A5-0197/2003. European Parliament, Strasbourg.

Fischler, F. (2001) *Choosing our Policy - European Agriculture Meeting the Challenges of WTO and Mid-term Review.* Confederation of European Agriculture Conference, 26 September 2001. CEA, Belfast.

Fischler, F. (2003) *WTO and agriculture.* Press Release IP/03/231, 13 February 2003. European Commission, Brussels.

Glavany, J. and Künast, R. (2001) La réforme en continu. *Le Monde*, 31 Juillet 2001.

Haniotis, T. (2002) *The New US Farm Bill From an EU Perspective.* Paper presented to the 78th Congress of the French Wheat Growers' Association, Blois, 6 June 2002.

Massot, A. (2001) *La Multifuncionalidad Agraria, un Nuevo Paradigma a la Búsqueda de una Política Común en la Era de la Globalización*, IV Congreso de la Asociación Española de Economía Agraria, Madrid.

MADRP (2003) *Impacto na Agricultura Portuguesa da Proposta da Comissão em matéria de Desligamento das Ajudas Directas.* GPPA, Lisboa.

MAPA (2003) *Efectos que la Propuesta de Reforma de la PAC Tendrá Sobre la Producción de los Cultivos COP, las Leguminosas y el Arroz.* DGA, Madrid.

OECD (2001a) *Agricultural Policies in OECD Countries - Monitoring and Evaluation.* OECD, Paris.

OECD (2001b) *Agricultural Policies in OECD Countries - a Positive Reform Agenda.* OECD, Paris.

Ruffer, T. and Swinbank, A. (2003), *Stock-take of the WTO Agriculture Negotiations. Implications for Developing Countries.* Oxford Policy Management, Oxford, UK.

World Trade Organization (2001) Ministerial Conference, Fourth Session, Doha 9-14 November 2001. *Ministerial Declaration.* WT/MIN(01)/DEC/W/1. WTO, Geneva, Switzerland.

World Trade Organization (2003) Committee on Agriculture, Special Session. *Negotiations on Agriculture. First Draft of Modalities for the Further Commitments. Revision.* TN/AG/W/1/Rev. 1. WTO, Geneva, Switzerland.

Chapter Ten:

Concluding Comments

Alan Swinbank and Richard Tranter
Centre for Agricultural Strategy, The University of Reading, UK

The policy background

We noted, in the Preface, the Commission's frank admission in early 2002 that the direct payments that had played a key role in *compensating* farmers for the reductions in intervention prices agreed in the 1992 MacSharry reforms had 'lost part of their compensatory character after ten years of implementation' and had 'instead become simple direct income payments' (Commission, 2002a). Given that this was the entrenched view of the majority of European farm leaders, and that it was firmly supported by France - and in particular the French President Jacques Chirac - there is little surprise that, in Copenhagen in December 2002, the European Council decided to extend the geographical coverage of the scheme to the new states of Central and Eastern Europe upon their accession to the EU in 2004. This was despite the half-hearted attempts of some Member States, in particular the UK and Germany, to re-open the debate on the purpose and future scope of direct payments, as outlined in Chapter 2. It was far too risky a political strategy to face the opprobrium a breakdown of the accession negotiations would trigger if they sought to make enlargement conditional upon CAP reform.

During 2002 the EU Farm Commissioner, Franz Fischler, embarked upon *his* reform of the CAP. As Arlindo Cunha makes clear in Chapter 9, Fischler's determination stemmed in part from what he, Fischler, perceived as the failure of the Agenda 2000 reform agreed in Berlin in 1999. It pitted Fischler against Chirac, who saw the Agenda 2000 reforms as a personal success that would carry the CAP through to 2006. With French presidential and parliamentary elections due in 2002 - Chirac regained the presidency and the centre-right replaced the socialists as the governing party - the Commission was not in a position to present its proposals for a *radical* mid-term review until July (Commission, 2002b). When it did present its reform proposals these contained a key component - the leitmotiv of the package according to Cunha. This was for a further decoupling of direct payments;

decoupling them from production, but retaining the link with farmland. Although this proposal fell short of the bond scheme advocated by Swinbank and Tangermann in Chapter 4, measured against the historical progression of the CAP it did represent, and was seen as, a *radical* proposal for reform.

Quite when Fischler determined upon the content of his reform agenda, or where the idea for decoupling came from, is unclear. Cunha's analysis suggests that among Fischler's key aides, if not in the Commissioner's mind, the outline of a radical reform was taking shape in 2001 or earlier. Future autobiographies of key participants, and structured interviews with Fischler's inner policy-making circle, will doubtless elucidate this question in years to come, in much the same fashion that the *process* of the MacSharry reforms has been the subject of continuing analysis. However, Cunha's suggestion that the Commissioner was reluctant to reveal his hand until the French election cycle had been completed is credible.

But where did the idea come from? As both Daugbjerg (in Chapter 6) and Cunha (in Chapter 9) make clear, the Commission has a powerful role to play in defining the scope and content of reform: other actors would face considerable difficulty in injecting new ideas into the policy-making process, as the focus of the Council's discussion is, inevitably, the Commission draft.

We have shown, in Chapter 3, that advocacy of decoupled compensation payments has a long history in the academic literature, dating back to at least the early 1960s (e.g. Nash, 1960). In 1990 this academic work resulted in a proposal for a bond scheme in a report commissioned by the Land Use and Food Policy Inter-Group (LUFPIG) of the European Parliament (Tangermann, 1990 and 1991). One suspects that it was *this* study which either consciously or subconsciously prompted the Danish Minister's advocacy of a bond scheme for arable crops during the discussions on the MacSharry reforms.

However, in 1990-1992, in addition to being at variance with the Commission's proposals for CAP reform (even though the Commission was itself proposing a bond scheme for milk producers), the LUFPIG and Danish proposals were both too *new* and too *radical* as far as CAP policy-makers were concerned, and failed to command support.

Nonetheless, the MacSharry reforms could be seen as a first step towards a bond scheme. By partially breaking the link with production and determining a payment entitlement based on past production decisions, and in determining the amount per hectare or per animal of what was then said to be a compensation payment, the policy distance between the pre-1992 CAP and the bond scheme was considerably diminished. Whether Mr MacSharry foresaw a further evolution of the scheme is unclear. But, nearly a decade later, it may be that Mr Fischler and his close colleagues, in deliberating on the scope and form of the mid-term review, saw a further decoupling of support as a logical extension of the MacSharry reforms. Set in the context of a WTO system that favoured decoupled - green box - measures, various OECD studies on decoupling, and a general political climate in which discussants frequently referred to 'decoupling', even though they might have had an imperfect understanding of the concept, this would be unsurprising.

The Commission Services, nonetheless, were aware of the bond scheme. The Framework V project that led to this book was approved by DG Research in 1999; and the Premier Issue of *EuroChoices*, that contained Swinbank and Tangermann's advocacy of a bond scheme (Swinbank and Tangermann, 2001) also carried a 'welcome' from Fischler in which he stated:

> 'To help us confront the major challenges facing European agriculture, the exchange of information and ideas, as well as in-depth economic analysis and debate, is needed more than ever before. The Commission recognised these signals at an early stage and to explore the increasing complexity of agricultural policy issues is actively supporting the work of researchers and institutions, through research programmes and the involvement of our own personnel' (Fischler, 2001).

It is likely that someone in Mr Fischler's *cabinet*, if not the Commissioner himself, was aware of the bond scheme proposal.

If the Commissioner, or his staff, was aware of the bond scheme proposal, did this encourage or discourage the Commissioner's appetite for radical reform? We must await the verdict of future studies to answer this question. On the one hand the Commission had emphasised the *economic* benefits of decoupling, whilst continuing to stress agriculture's multifunctional role. On the other hand, the Commissioner will have been very well aware of the *political* difficulties he would face in proposing a radical CAP reform.

The economics of decoupling

The main economic *benefit* of decoupling is that it gives farmers greater freedom to farm, thereby potentially reducing their costs and increasing their profits. For the economy as a whole, this means an increase in welfare, as agricultural output is produced at lower cost. As European farmers become more competitive internationally, this would enable the EU to agree to more stringent controls in the WTO on the design and extent of farm support programmes. The switch from the blue to the green box categorisation, as a result of the implementation of the mid-term review, is a practical expression of this change, although it may not be sufficient for a successful conclusion of the Doha Round.

In presenting the mid-term review package to the British House of Commons, the Secretary of State spoke of 'breaking the link between farm subsidies in order to reconnect farmers to their markets, reduce damaging environmental impacts and reduce bureaucracy' (Beckett, 2003). Here she is choosing to emphasise the negative externalities associated with agricultural production. Hence, a decrease in agricultural output will reap an additional economic benefit: a reduction in 'damaging environmental impacts'.

Despite this claim for environmental gain, most of the *European* debate on the effects of decoupling focuses instead on the positive externalities associated with agriculture - its multifunctional role. The purpose of this study was not to address directly the issues associated with multifunctionality. But in Chapter 5, the Portuguese members of the team do set out how a bond scheme fits within a transition from an old-style CAP to something akin to the common agricultural and rural policy for Europe (CARPE) advocated by Buckwell *et al.* (1997).

The issue centres on the fear that with a decline in agricultural production, the viability of rural communities will be endangered (leading to 'desertification'), the visual attraction of the countryside will decline, and fauna and flora associated with particular farming systems will be lost. It is probably fair to say that both views - on the negative and positive externalities of agriculture - are reconcilable. One tends to focus on the intensive margin, where the drive to increase production is seen to result in an adverse environmental impact; the other focuses on the extensive margin, where more 'traditional' farming methods are practised on more marginal land. Overall, our sample of farmers was remarkably sanguine when asked how they might respond to the introduction of decoupled payments, with some 70% in each of the three study countries responding that they would not alter their mix of farm activities and at least 50% of those in each of the three countries responding that they would leave no land at all idle.

The economics of multifunctionality are complex, and whilst much thought has been given to the topic in recent years, more needs to be done to reconcile divergent views and enhance the pool of empirical evidence on which policies need to be based. The issue centres on the fact that both positive and negative externalities - non-marketed goods or services which do nonetheless confer costs or benefits on individuals who are not party to the prime market transaction - are *joint* products with farm production. The question then is: what degree of jointness is involved? If the two can be separated by the use of appropriate technologies or management techniques, if for example pigs can be reared without the creation of obnoxious smells, or if hedgerows can be maintained without animals being grazed in fields, then the advocates of free, or freer, trade will suggest that farm production should respond to world market signals, with an appropriate set of taxes and subsidies in place to elicit the appropriate, *economic*, supply of pollution or environmental goods. These are not easy issues with which to grapple, but it does leave us firmly in the realm of environmental economics (see, for example, Pearce and Turner, 1990).

If, on the other hand, the jointness is complete, that there is no way in which the 'bad' externality can be reduced or the 'good' externality can be increased without engineering a comparable change in farm output, a different scenario applies. A policy focussed on farm production (e.g. free trade) will simultaneously determine the supply of the externality. In practice, most situations will lie between these two extremes, and may well be site specific. The challenge then becomes that of devising workable schemes that reduce bureaucracy (transactions costs) to acceptable levels, whilst delivering benefits valued by society (at a cost acceptable to the Treasury) and matching the green box specifications of the WTO (Swinbank, 2002).

In Chapter 4 it was asserted that a further advantage of a bond scheme was that it would remove the policy uncertainty attached to the current regime. Whilst our farmer respondents disagreed with the statement that 'the current arrangements under the CAP work well', our supposition was not endorsed by the finding from the postal survey that farmers in each of the three study countries were in strong agreement that the current pattern of support is a reliable source of income for farmers.

The politics of decoupling

For a CAP reform to be agreed, it needs to proceed through a number of formal steps:

1. It must be proposed by the Commission, which implies the Commissioner with the support of the College of Commissioners;
2. It must receive the support of France, but also Germany, despite the provisions for qualified majority voting in the Council of Ministers; and
3. It needs a qualified majority vote in the Council of Ministers.

As yet, the European Parliament has a limited role in this process, but that may change in the future.

Conditioning the views, and limiting the freedom of manoeuvre of the main participants (the Commissioner, and the 15 (soon to be 25) Farm Ministers), are international (e.g. WTO) constraints, budget considerations, and the reactions of the main domestic stakeholders who will be impacted by the reform. Clearly all these factors interplayed in the crafting of the partially decoupled Single Payment Scheme introduced into the CAP in the 2003 Fischler reform. Fischler's tenacity was crucial. Would a future Commissioner be willing to champion a bond scheme?

In addition to concerns about the loss of agricultural production (and hence its multifunctionality) in marginal areas, in stakeholder discussions farmers were said to be concerned about the longer-term political acceptability of being paid to do nothing, and of unfair competition between subsidised and unsubsidised farmers.

In the mid-term review, the EU has attempted to address the first concern by tying the Single Payment Scheme to land that remains in agricultural production, by imposing cross-compliance, and through an increased funding of Pillar 2 activities. The experience of implementing these provisions will give policy makers useful information on which to base future reforms, but this implies a stringent programme of monitoring and evaluation. A bond scheme, of course, would be completely decoupled from land use, and thus cross-compliance would not be an option.

By linking payments to 'farming' in this way, if not to farm production, the EU has addressed the concern that farmers will be seen to be paid to do nothing. Indeed, the Commissioner claimed that the outcome of the mid-term review would 'help improve the public image of and support for the common agricultural policy' (Fischler, 2003). Thus, farmers will receive a Single Payment, with conditions (cross-compliance) attached, but with no obvious termination date. By contrast, with a bond scheme, Swinbank and Tangermann (2001) envisaged a series of *compensation* payments over a clearly defined time horizon. The farmers in our survey, however, showed few qualms about their willingness to accept the bond scheme, with 45% of the British sample, 35% of the German sample, and 69% of the Portuguese sample saying that they would be very or quite comfortable at the thought of receiving payments unattached to production under the first two steps of the bond scheme (with 23%, 45% and 15% respectively saying they would be quite or very uncomfortable, and the remainder expressing no preference).

As a system of farm income support, the Single Payment Scheme does not address the equity concern. Payments on an historic basis will still derive from the farm's past enterprise mix and scale of production. True, payments in excess of €5,000 will be subject to modulation, with the funds redeployed to the Second Pillar, but with very limited redistribution between Member States. But, it cannot be claimed that the payments are related to need, or that they are directly linked to the delivery of environmental goods. Furthermore, the beneficiaries will be the present generation of farmers (and landowners), for the value of the Single Payment will tend to be capitalised into land values (and rents), raising entry costs into farming and potentially leading to a lock-in of policy. By contrast, the bond scheme advocated in Chapter 4 would make a clean break, admittedly *compensating* past recipients of direct payments, but with that legacy fading as compensation payments were phased out.

CAP reform 2012?

One prediction that can be safely made is that two or three years hence, with a new Commissioner in office and international pressures unabated, pundits will once again debate the possibility of further CAP reform. The MacSharry reforms of 1992 started a process of compensation through decoupling. During the 1990s the focus on compensation was lost, but the Fischler reforms of 2003 led to a further decoupling of direct payments. Some commodity regimes, such as sugar, remain unreformed, but it seems inevitable that the MacSharry-Fischler package will be brought to bear across all commodity regimes during the next ten years.

The unanswered question is whether the CAP of the 2010s will be based on the Single Payment Scheme with the Second Pillar playing a subsidiary role, as now; or whether Single Payments are themselves phased out, with a new CAP becoming focussed on protection of the environment and promoting rural development but also encompassing some price and income stabilisation mechanisms. A bond scheme would facilitate that switch, and we hope that this book has helped identify some of the issues that would then need to be addressed by policy makers.

References

Beckett, M. (2003) Secretary of State Margaret Beckett's statement to the House of Commons on CAP reform. 1.15pm, 26 June 2003. Department for Food and Rural Affairs. www.defra.gov.uk/corporate/ministers/statements/mb030626.htm (accessed 11 August 2003).

Buckwell, A., Blom, J., Commins, P., Hervieu, B., Hofreither, M., von Meyer, H., Rabinowicz, E., Sotte, F. and Sumpsi Vîna, J.M. (1997) Towards a common agricultural and rural policy for Europe. *European Economy* Reports and Studies No 5. CEC, Luxembourg.

Commission of the European Communities (2002a) *Enlargement and Agriculture: Successfully Integrating the New Member States into the CAP. Issues Paper.* SEC(2002)95. CEC, Brussels.

Commission of the European Communities (2002b) *Communication From the Commission to the Council and the European Parliament. Mid-term Review of the Common Agricultural Policy.* COM(2002)394. CEC, Brussels.

Fischler, F. (2001) Commissioner Fischler Welcomes EuroChoices. *EuroChoices* Premier Issue, Spring, 5.

Fischler, F. (2003) CAP Reform. Speech to the Committee on Agriculture and Rural Development of the European Parliament, Brussels, 9 July 2003. Speech 03/356.

Nash, E.F. (1960) Farming with free prices: towards a competitive agriculture. *The Guardian*, 18 August.

Pearce, D.W. and Turner, R.K. (1990) *Economics of Natural Resources and the Environment.* Harvester Wheatsheaf, Hemel Hempstead, UK.

Swinbank, A. (2002) Multifunctionality: the concept and its international acceptability. *Journal of the Royal Agricultural Society of England* 163, 141-148.

Swinbank, A. and Tangermann, S. (2001) The future of direct payments under the CAP: a proposal. *EuroChoices* Premier Issue, Spring, 28-29, 32-34.

Tangermann, S. (1990) A bond scheme for supporting farm incomes. In: Marsh, J.S., Green, B., Kearney, B., Mahé, L., Tangermann, S. and Tarditi, S. (eds) *A Future for Europe's Farmers and the Countryside.* Land Use and Food Policy Inter-Group. European Parliament, Brussels.

Tangermann, S. (1991) A bond scheme for supporting farm incomes. In:Marsh, J.S., Green, B., Kearney, B., Mahé, L., Tangermann, S. and Tarditi, S. (eds) *The Changing Role of the Common Agricultural Policy: the Future of Farming in Europe.* Belhaven, London.

Index